The Zookeeper's Daughter

Lori Space Day

PublishAmerica
Baltimore

© 2004 by Lori Space Day.
All rights reserved. No part of this book may be reproduced, stored in a retrieval system or transmitted in any form or by any means without the prior written permission of the publishers, except by a reviewer who may quote brief passages in a review to be printed in a newspaper, magazine or journal.

First printing

ISBN: 1-4137-3260-7
PUBLISHED BY PUBLISHAMERICA, LLLP
www.publishamerica.com
Baltimore

Printed in the United States of America

This book is dedicated to the women of Space Farms, whose unheralded work behind the scenes has enabled the farm to thrive.

Many thanks are extended to Karen Talasco, Donna Traylor, Michael O'Sullivan and Dr. Leonard Lee Rue III for their photographic contributions to *The Zookeeper's Daughter*.

I would like to thank my mechanical genius, computer guru husband, Doug. Without his patient tutoring and active participation this book would not be possible.

I would also like to sincerely thank the animal mentors in my life: my grandmother, Elizabeth; grandfather, Ralph; my other grandmother, Elinor; my Dad, Fred; my brother, Eric; and family friends: Joe Taylor, Dr. Leonard Lee Rue III and Al Rix. I have learned so much from you. You are truly caretakers of the earth and its creatures. It is my hope that the generations that come after us will learn, through our efforts to educate, that God gave mankind dominion over the earth and its creatures to observe, protect, rejoice and enjoy.

The Spaces
—By Dr. Leonard Lee Rue III

It was in the summer of 1936. I was ten years old and it was a day I never forgot.

My father had bought a farm just north of Belvidere, NJ in 1935, but we hadn't moved up there as yet. We lived in Paterson and went up to the farm every weekend. I have always been fascinated with wildlife and, on the farm, I was in the fields or woods all of the time. I had started writing my observations of wildlife when I was eight years old.

On the way up to the farm one weekend, My Uncle Tom turned left off Route 46 in the Parsippany-Troy Hills area. He wanted to stop at the Morris County Fair and, of course, I did, too. We looked at a number of exhibits and for the life of me I couldn't tell you now what they were. Then he took me into a world of its own, my kind of world. It was Ralph Space's exhibit of the wildlife of New Jersey. Ralph had red and gray foxes, a bobcat, raccoons, opossums, mink, a groundhog and a couple of owls. I had seen groundhogs; we had lots of them on the farm. I had seen raccoons and opossums crossing the road at night. I had never seen a fox or a mink and certainly not a bobcat! I was fascinated.

Ralph Space had snakes, real live timber rattlesnakes and copperheads. I didn't even know there were such snakes in New Jersey. I was mesmerized.

I actually got to talk to Ralph Space, or rather he took the time to talk to me. He told me about wildlife, something no one had ever done before.

He actually knew all about wildlife. I had never met anyone like that before. It was only years later that I really knew, and appreciated, how much he really knew about wildlife. I knew Ralph never remembered me; I was just one of the thousands upon thousands of kids that he had told about wildlife, but I never forgot him.

When I was seventeen, I met Joe Taylor and Leon Kitchen, two of the country's top fox trappers. By that time, I was beginning to make a name for myself as a fox trapper, too, and I actually trapped in competition with both of them. From the start, Joe and I became very close friends and he was my mentor in everything pertaining to wildlife. Not only that but he was practically part of Ralph Space's family. After he had taken me up to Space Farms a number of times, I became a member of that select family group, too. I'm just a few years older than Fred and he and I became very close friends. Fred has always been one of the hardest working men I ever knew.

By the time I was nineteen, I had embarked on a career of being a professional wildlife photographer and have since become the most published wildlife photographer in the country. The Spaces, all of them, with their large stock of wildlife at Space Farms have always been and still are, ready to help me if they have anything that I need to photograph. Years ago, Ralph and Fred told me that I could take out anything that I could get back in.

I have taken out and photographed bear cubs, lion cubs, wolf pups, woodchucks, badgers, owls etc. Fred has cut holes in the fence so I could photograph the African lions, tigers, bears etc. I have worked in the snake pit and photographed dozens of snakes that we have taken out. I once built a mock-up log cabin inside their wolverine pen so I could get photographs for *True* magazine, back when it was "the man's magazine." I have photographed their deer and have gotten deer from their herd to add new bloodlines to my own herd.

I have eaten dozens of buffalo burgers in their restaurant, drank a lot of tea and eaten lots of cookies in Mom's kitchen. I was amazed when Elinor showed me how she had trained a black bear cub to sit on a potty after she had fed it, and how that cub didn't want to go out and be with the "animals" when he got too big for the apartment. I remember when Parker, as a baby, slept in the playpen in the restaurant with the baby grizzlies; they all got their bottles at the same time.

Eric has carried on the legacy of being an animal person and is a top nuisance animal control man. As for Lori, the author of this book, I knew her before she was Lori.

I have known, worked with and loved three generations of Spaces. It has been an honor to be part of their extended family and it's been a privilege to write this forward for Lori's book.

Table of Contents

Prologue..13
The Early Years
Chapter 1   Maggie the Black Bear...............................17
Chapter 2   Buzzy the Turkey Buzzard........................19
Chapter 3   Jimmy the Crow...21
Chapter 4   Horseback Riding with Gramp..................24
Chapter 5   Johnny the Porcupine................................25
Chapter 6   Isabel the Mouflon Sheep..........................28
Chapter 7   Raccoons and Kids.....................................31
Chapter 8   White Tail Deer Fawns..............................32
Chapter 9   Vacations and Holidays.............................33
Chapter 10   Salamander Hunts......................................37
Chapter 11   Boris the Otter..38
Chapter 12   Bonnie and Benny, the Chimps.................41
Chapter 13   Skunks...43
Chapter 14   Jingles the Jaguar.......................................45
Chapter 15   Bear Cubs...49
Chapter 16   Moses the African Atlas Lion..................51
Chapter 17   Night Capers..56
Chapter 18   Geronimo and Little Breeze, the Cougars....58
The Pet Shop Years
Chapter 19   The Pet Shop Years...................................62
Chapter 20   Jacquelyn Renee..67
Chapter 21   Meanwhile, Back at the Farm...................70
Home Agin'
Chapter 22   Belly Rub, the Coyote...............................73
Chapter 23   Kiwi the Coatimundi.................................75
Chapter 24   Henrietta the Chicken................................78
Chapter 25   Snakes on the Road....................................80
Chapter 26   Winking, Blinking and Nod, the Kestrels....83
Chapter 27   Syrian Grizzly Bears..................................86
Chapter 28   Monkey Business.......................................90

Chapter 29	Moving Day	92
Chapter 30	Wild Turkeys	99
Chapter 31	Three Little African Atlas Lions	104
Chapter 32	Guns and Brass	107
Chapter 33	The Leopard's Eye	109
Chapter 34	Norman the Holstein Calf	111
Chapter 35	Three Little Stinkers	113
Chapter 36	Dead Snake on the Road	117
Chapter 37	Yankee Doodle Dandy, the Lion	118
Chapter 38	Salutations of Holiday Cheer: A Note to my Friends	126
Chapter 39	The Lion's Tooth	128
Chapter 40	Blizzard and Cousins	132
Chapter 41	Jimmy the Llama in Pajamas	141
Chapter 42	Gramma's Snowflake	148
Chapter 43	Solomon, a New King for our Pride	151
Chapter 44	Gramma	153
Chapter 45	This Little (Piggy)	159
Chapter 46	Inca Macaw	161
Chapter 47	Oh Deer, a Broken Leg	163
Chapter 48	Santa May be Late This Year	170
Chapter 49	Pebbles, Brook and Rocky, the Fishers	172
Chapter 50	Ivy Lion Takes a Road Trip	175
Chapter 51	Pickles the Opossum	178
Chapter 52	Niles and Sahara, the African Atlas Lions	181
Chapter 53	The Anti-condom Snake	183
Chapter 54	Animal Photography	185
Chapter 55	What's in a Name?	188
Chapter 56	Tara and Khyber, the Tigers	190
Chapter 57	Happy Holidays	201
Chapter 58	Lizzie the River Otter	203
Chapter 59	Chip and Dale, the Flying Squirrels	214
Chapter 60	Baby Beavers	218
Chapter 61	Tara and Khyber Graduate to the Zoo	223
Chapter 62	Hurricane Floyd	226
Chapter 63	Sweet Pea, Cowboy, Chiquita, Turk, Dale and Annie, the Lions	229
Chapter 64	Y2K Bug Holiday Letter	233

Chapter 65	Painting in the Zoo	235
Chapter 66	Lucky the Fawn	239
Chapter 67	Bloomfield Boomerang, the Hawk	243
Chapter 68	A Soused Boar in New York City	246
Chapter 69	A Moose and my Caboose	249
Chapter 70	Duck Ale Grill	253
Chapter 71	My First Buck	256
Chapter 72	Tiger Tara's Tail Tale	259
Chapter 73	Christmas Letter 2000	263
Chapter 74	Bottle Breaking Multiple Species	265
Chapter 75	Rescuing Jane Doe	272
Chapter 76	Three Inquisitive Raccoons	274
Chapter 77	Fritzy the Black Australian Swan	278
Chapter 78	Pushing the Buck	282
Chapter 79	Hope and Hannah, African Atlas Lions	286
Chapter 80	Instinct	289
Chapter 81	Swimming with the Dolphins	292
Chapter 82	Holiday Greetings 2001	295
Chapter 83	The Lions Finally Catch On	298
Chapter 84	Little Boy Blue in the Nursery	302
Chapter 85	Angie, the Molluccan Cockatoo	306
Chapter 86	The Boa of Stillwater	312
Chapter 87	Kernel Corn Does Not Grow Tomatoes	315
Chapter 88	Holiday Time	318
Chapter 89	Goodness Gracious, Great Balls of Fur	321
Chapter 90	Angie Can Fly!	330
Chapter 91	Uh Oh, Snake's Loose	333
Chapter 92	Nine O'Muck and Mire	338
Chapter 93	G.I. Gator	342
Chapter 94	Christmas Pearls	350
Chapter 95	Dear Jackie	354

Prologue

Space Farms Zoo and Museum began in 1927 when Ralph and Elizabeth Space bought the first 1/4-acre at the same site of today's 100-acre complex. The Spaces were native farm people, both growing up on small family farms in Sussex County, NJ.

Ralph Space worked hard to support what was soon a growing family. Ralph's mechanical expertise enticed him to open a small repair shop and gas station. Elizabeth helped by starting a small general store carrying the few necessities the local people needed: salt, corn flakes, bulk cookies, soda pop, flour, and other dry goods like Levis, gloves, socks, etc.

Later to supplement the family income, Ralph, a hunter and trapper since childhood, was employed by the state of New Jersey Game Department to trap predators (varmints), which attacked local farm animals. This moonlighting was natural for Ralph, who was always an avid sportsman.

Most of the distress calls were from local farmers in the springtime, when native bobcats, foxes and raccoons had young to feed and preyed on the plentiful farm goats, sheep, cows, chickens and ducks. Rather than kill the animals in the spring, when the fur was of little value, Ralph built small enclosures around his garage to keep them till fall.

When the time came, however, the three small Space children, Loretta, Edna and Fred cried and begged their father to keep them. By next spring the animals had multiplied and the collection of wildlife grew. Ralph realized the potential of ranch fur farming, continuing to breed different types of foxes, and imported the first pair of Canadian platinum foxes to the United States. In 1933 Ralph Space had a mink and fox ranch and had started the wild animal farm.

People around the area soon heard about the Space family's wild animal collection and would drive by to see it. They would buy gas, soda, and maybe some penny candy and visit the animal collection behind the garage. Ralph Space had a small zoo.

Neighbors would often bring injured or orphan wildlife to the gas station for the Spaces to care for, because "Those Spaces know all about animals." The zoo grew. Soon bears, white tail deer, bobcats and porcupines joined the bobcats, foxes, raccoons, mink, squirrels, skunks (deodorized of course!) and woodchucks in the enclosures behind the general store and gas station. In 1933 Ralph obtained a license for his wild animal farm from Wantage Township at the cost of $1.

Soon after the zoo began behind the gas station, the Great Depression hit the country. The local farmers were not spared. Still needing supplies and repairs the farmers and neighbors would "pay" with their family heirlooms. Old firearms, dolls, cradles, dishes, treasured original farm inventions, butter churns etc. hung on the walls in the general store until their owners could afford to pay their bills. Sometimes the owners came back and claimed their goods, sometimes they did not. The Space Wild Animal Farm Museum had very humble beginnings.

Visitors coming for gas would stop and see all the items and animals. One day a man mentioned to Ralph, "This place is great, you ought to charge admission!" Ralph swooped off his hat and said, "5 cents please!" And so Space Farms Zoo and Museum became an official tourist attraction.

As time went on, the children worked diligently. Traditionally all farm children worked the farm with their parents. Fred Space's chores as a child became his calling as an adult. Under the watchful eyes of his parents he learned the intricacies of wild animal care.

In 1951 Fred married Beverly Nankivell, a young beauty from a local lake community. Her family had a summer home where Fred had done bulldozer excavating. With a growing family, Fred took a keen interest in building up the zoo. Under his guidance the zoo branched out into exotic animals in addition to the native wildlife.

Today Fred Space is a renowned naturalist not only in the United States, but worldwide. He is responsible for an extensive genetic diversity breeding program that helped to replace Hokkaido bears in Japan, Elk in South Korea, Fallow deer in Taiwan, and helped to re-establish Wild Turkeys throughout the New York, New Jersey, Pennsylvania area. Under Fred's care the Space Farms Zoo has established longevity records for the Bobcat, Jaguar and Cougar, and was home to the Guinness World Record Kodiak bear—Goliath. Goliath weighed in at 2000 pounds! Goliath lived to be equal in age to a 96-year-old man. Up until Goliath's death, Fred Space was able to feed the behemoth bear marshmallows by hand.

Loretta and Edna left the farm after marrying, however, can often be found in the restaurant chatting with family, neighbors and friends.

Grampa Ralph's enthusiasm for collecting continued until his death in

THE ZOOKEEPER'S DAUGHTER

1986. His legacy lives on in a phenomenal early Americana Museum complex housed in nine refurbished barns. Gramma Elizabeth Space worked in the store right through its transition into the current restaurant. She "retired" at age 89, but still supervised the restaurant until she passed away in 1997.

Today the third generation of Spaces, Fred's children, help Fred run the zoo. That's my brother, his wife, my husband, and me...

I am the zookeeper's daughter.

Chapter 1
Maggie the Black Bear

According to the family, as I was too young to remember this story… I loved to play, as all kids do, with animals. Most kids have a cat or a dog. The Spaces had received an orphaned black bear cub. It was raised on a bottle in 1956, with the formula developed by my Gramma for infant bears. The bear cub's name was Maggie. Black Bear cubs are born in January and are about 8 ounces at birth. Born with eyes closed and hairless, they are quite fetal in appearance. They grow rapidly and by mid summer they are about the size of a medium sized dog, and as tall as a two-year-old child is. Raised by hand they are friendly, but still play with teeth and claws. That did not daunt a two-year-old child. Maggie was my pal. It is tradition in my family that when it comes time for a human child to give up the bottle, the bottles are "given" to an animal baby that "needs the bottle more than you do." I had given up my bottles for Maggie. She loved me and I loved her. We took baths and played together. My sister Renee has pictures of us in the bathtub with baby bear cubs.

Every day my dad and mom, my older brother, Eric, and younger sister, Renee, and I would have lunch in the general store with my Gramma. Mom worked in the general store, so it just made sense to eat there. One day lunchtime came and I was nowhere to be found. I had snuck away. Everyone looked for me, all over the place. There are lots of places for a small two-year-old to hide in a general store. My parents were panic stricken, as all parents are when their child is lost. The panic is worse when there are dangerous animals around. My dad went outside to look for me. After searching the grounds, he spotted me. I had unlocked the gate and let Maggie out to play. Bears play rough and tumble, so Maggie and I took turns running and wrestling. Dad was concerned. He knew if he tried to break up our rough and tumble play, Maggie would play harder, possibly hurting me. If I did not come willingly, and cried, Maggie would protect me, and turn on him. He had to get me out without upsetting my bear buddy, so my dad, smart father that he was, gently called to me, "Lori, time for ice cream!" I came running. As I came out the gate, with Maggie on my heels, Dad called me to the kitchen once again. I ran into the kitchen through the screen door. Dad quickly slammed the screen door, leaving Maggie outside, whining for her lost friend. Dad went outside once again and gave Maggie ice cream, (a favorite treat for bears), to entice her back into her cage. That year all the cages were padlocked. I had to eat a good lunch before I got my ice cream, but I was never punished for loving an animal.

Chapter 2
Buzzy the Turkey Buzzard

When I was 4 years old Dad raised a young turkey buzzard by hand. It was only the size of an orange, but as animal kingdom babies go, really ugly. In 1965 a fad started up with troll dolls. (Remember them?) Baby turkey buzzards are that kind of ugly. Ugly like only a parent could love. Turkey Buzzard babies are all white fuzz with a naked gray and pink skin covered head, large beady eyes, a beak that overpowers the face, and a knobby protruding neck crop (the first stop in the digestive system of a bird).

Adult turkey buzzards are no beauty contest winners either. Their red skin covered head has a purpose: turkey buzzards eat carrion—dead, often rotten, dead animals. The skin on the head cleans easier than feathers would. Turkey buzzards are large black birds, 25 to 34 inches long with a wingspan of 6 feet. They fly with a V shape to their wings, making them very recognizable against the blue sky over the local farmland. They are not killers; they are only scavengers and often seen circling above dead animals on local farms. The Cherokee Nation calls them "peace eagle." They are an ugly, but noble bird.

Buzzy stayed with Dad wherever he went, perched on his shoulder. Dad was so proud of his baby, feeding him scraps of meat from his pocket. Zoo visitors were enthralled by this ugly bird. Buzzy was sweet, kind, and used to human contact. One day, a lady visiting the zoo said to Dad, "What is that ugly bird?" My Dad was highly insulted. Buzzy was a member of our family. We loved Buzzy. Dad threw the lady out of the zoo. I don't think she ever returned—she must not have been an animal lover.

Dad, Fred Space and Buzzy's grandchild.

Chapter 3
Jimmy the Crow

My Gramma was always raising infant animals. It was not unusual for her to have fawns, mink kits or young raccoons in a box next to the stove in the general store to keep warm. (We are country people, so I feel it is important to point out the infants were kept next to the stove, not in a pot on top of the stove!) Infants had to be kept warm and the oven was the warmest place, plus she could keep an eye on them. Gramma also told me of keeping my dad literally in the oven to keep him warm as he was 3 pounds at birth. Gramma was a noteworthy cook and was always in the kitchen preparing meals for visitors to the zoo or whipping up special formulas for infant animals. Those were the days before the board of health was even invented, of course. (Germs never bothered farm kids; we grew up with them.)

I grew up surrounded by animals of all species. My Gramma and Dad would keep an eye on me and instruct me how to be safe around the animals. This is comparable to how a city parent teaches their child how to cross the street. I learned at my gramma's side and my dad's knee how to stay safe around the animals. "Hands on education" is the technical term today.

When I was 6 or 7 years old, Gramma raised a crow named Jimmy from a fluff ball about the size of a walnut. We would feed Jimmy tiny bits of meat on a toothpick. Jimmy became a favorite pet. When Jimmy grew big enough to be released, we put him in a lilac bush outside the admissions door. He would fly around but always came back. Jimmy liked to collect shiny things as most crows do. He would also share his treasures. Well, at least he would let me look at them. Brightly colored items would adorn his lilac bush. Jimmy also learned to say "Hellllllloooow," cat call whistle and laugh. By the way, you do not have to split a crow's tongue to enable him talk—that is an old wives' tale.

One day my dad was pulling weeds at the base of Jimmy's lilac bush. A handsome man came through the admission door, his arm possessively draped around a curvaceous blonde. Jimmy was our unofficial greeter. Jimmy whistled and cried out, "Helllllooow." The stranger did not think twice and punched my dad right in the eye! I can still hear Jimmy laughing.

Jimmy caused lots of trouble. He would swoop on unsuspecting visitors and steal their picnic sandwiches and cookies. And since crows love shiny objects, he often swooped on unsuspecting visitors and pecked at shiny jewelry.

As long as I can remember, Space Farms always had a Snake Den. Gramp, Dad, or my brother Eric would answer distress calls from neighbors and go pick up offending snakes on the neighbors' property. It was considered a duty we performed to protect the native snakes from being killed by uneducated and fearful neighbors. The men would bring the snake back and put it in the snake den for the summer. Come fall, the men would take the snakes back to the secret mountain dens only my family knew about. We borrowed snakes from nature, fed them better than they would have eaten in the wild and returned them every fall so they could hibernate naturally. Constrictors and poisonous snakes were all treated with the dignity they deserved. My Gramp Ralph, (and later my dad, brothers and myself), would stand inside the snake den and give educational lectures on snakes.

Snakes make a lot of people very nervous only because they do not know enough about them. Nonetheless, most people don't care for them. One day, Dad was in the famous Space Farms Snake Den giving his snake lecture. All of a sudden a lady screamed and fainted. While watching the snake lecture, Jimmy the crow had swooped down to peck at the lady's vibrant red toenails, so invitingly displayed in her open-toed shoes. After she came to, she swore a snake had bitten her toe! With all the commotion, smarty-pants Jimmy had flown back to his lilac bush. He sat there laughing. The next day Dad built a cage for Jimmy. He still talked, laughed and hollered, "Helllllooow" whenever someone came through the admissions door, but he pestered people "Nevermore."

Chapter 4
Horseback Riding with Gramp

Gramp kept a herd of saddle horses as long as I can remember. Gramp's prize horse was a palomino called Buckskin. Beer Belly, Skyrocket, Chestnut, Goldie and Blaze are some of the horses I remember. Gramp kept about a dozen horses in the field by the pond that is now part of the zoo. Every weekend throughout my childhood Gramp would have the ladies pack a picnic lunch for the grandkids and we would all ride horses up to the Appalachian Trail on the back end of the Space Farms property. Sometimes Dad came along; sometimes he stayed at the farm to get work done. We would follow the trail to the Look Out on Sunrise Mountain.

Sunrise Mountain faces east on the Blue Ridge of the Kittatiny Mountains. The Mountain was a revered spot for the family. It was a major rattlesnake den in the tri-state area. Sunrise got its name because it would "capture the sun," before the valley below, at dawn. It is this phenomenon that made it a warm spot for snakes to den, as the rock face would warm up quickly, giving the cold blooded reptiles energy for the day ahead.

As the youngest of the ten grandchildren that were allowed to go horseback riding, I rode with Gramp on his large palomino, Buckskin. Safely ensconced between Gramp's strong arms I rode through the woods to the trail. I rode with Gramp, in a position of honor. He called me "Sis." Gramp whispered all the secrets of nature and the woods in my ear as we went. I was special, a princess surveying my natural kingdom with my knightly escort. Gramp pointed out mosses and fungi growing on the trees, scat (poop) left by animals he would identify, rub marks on a tree from a buck deer's antlers, snakes sunbathing on rocks, raccoons sleeping in the crook of trees, cirrus clouds that looked like a horse's tail and many other natural wonders that surrounded us on our journey.

The other grandchildren followed behind us. The procession of horses and cousins would travel to the lookout and stop to have a picnic. Sunrise is a major geological landmark for migratory birds. We would eat as Gramp pointed out the birds, teaching us their calls and flight patterns. After lunch we would meander back to the farm, unsaddle the horses, put away the tack and brush down the horses. Those were memorable days.

Then one day the unthinkable happened. My younger sister Renee was to come along on the trail ride. She was going to ride with Gramp on Buckskin. I was heartbroken. And scared. I had never ridden by myself before. I was always safe with Gramp. I cried. I sobbed.

Whether or not Gramp knew I was heartbroken, I don't know. Maybe he just thought I was a chicken. He took me by the hand and told me, "Come on, Sis, you can do this, you know how, you just think you don't know how." That was a hard concept for a seven-year-old. Gramp led me to the horse. The horse was growing taller and bigger as I approached it. All the while Gramp kept repeating, "You can do it, Sis. You can do it." Gramp lifted me into the saddle and adjusted the stirrups to fit my short legs. I sat high on the saddle, and viewed the world from a new perspective. "You can do it." Gramp patted me on the thigh and mounted Buckskin. Dad handed little Renee up to my former position of honor. I never heard Gramp whisper to Renee or call her "Sis," though I'm sure he did, as he had done for all the other grandkids before me, and those who would come horseback riding after me. I never heard a word, because I was concentrating so hard to stay on that horse, because my Gramp knew I could do it. I didn't want to disappoint him, or be seen as a baby any longer. I was riding my own horse, enjoying the elation that only horsemen know. I may have been a little nervous, but I was alone on the horse. I could do it. Those words whispered in my ear as a child still ring loudly whenever I am in a tough situation. My mantra, "You can do it, Sis, you can do it," came from my Gramp.

Chapter 5
Johnny the Porcupine

Johnny was a North American Tree Porcupine. The tree porcupine is the common porcupine of the United States and Canada. Johnny was an Eastern Tree Porcupine. The porcupine is a large (size of a basketball), cumbersome nocturnal rodent. Its chubby rounded body is 18 to 23 inches long with an eight-inch tail. Like other rodents, the porcupine has large incisors. The porcupine uses these orange teeth to eat tree bark and leaves.

The porcupine is known for its quills. Approximately 30,000 quills cover the body of the tree porcupine. Each quill is 2 to 3 inches long, hollow, and has a barbed hook on each end. Porcupines use their quills for defense against predators. Lifting their quills, rolling their bodies into a tight ball, they, in effect, becoming living pincushions. Porcupines cannot shoot their quills; you have to touch the porcupine in order for the quills to stick in you. Once the quill is imbedded the barb keeps it hooked into the skin. The warmth of the aggressor, now victim, warms the air within the quill, causing it to expand, sending it farther into the muscle. The *American Journal of Mammalogy* cites a case where a quill fragment 3/4 inch long moved two inches in warm muscle tissue in twenty four hours. Many a farm dog has come home with quills in its mouth that must be pulled out with pliers, due to the barbs on the end of the quill. Without medical attention the results of a snout full of quills may be deadly, usually resulting from starvation or infection of a major organ.

The porcupine has few enemies; the fisher is the only animal known to successfully prey upon a porcupine. First the fisher scares the porcupine. The prickly pudgeball rolls up into a defensive position, tucking its head under its belly. The fisher then puts his paw under the stationary pincushion to its belly, the only area on the porcupine that does not have quills. Deftly, the fisher flips the porcupine over on its back and dines from the belly side out, avoiding the deadly quills. Porcupines have saved many starving people lost in the woods. They are easy to catch (since they do not run, but stay still and ball up), kill and the meat is tasty if you remember to go through the belly side of the animal. This unique animal was revered by many Native American cultures. The quills were often dyed and used as decoration for clothing and baskets.

Johnny the porcupine was special for us. Johnny had been delivered by Cesarean section by a trapper friend of my Gramp. Gramp's friend kept him for many years then finally decided to give him to our zoo. Johnny was unique, because he was delivered by Caesarian section, and was so used to human contact he would not stand his quills up in defense. Johnny was a petable porcupine as long as you remembered to pet him in only one direction—from head to tail. If you forgot once, petting the wrong way would result in a hand full of porcupine quills, and you would never forget again. The quills did not hurt too much until you had to pull them out! Petting Johnny was fun. Johnny loved chocolate and peppermint candy.

We Space kids gave lectures to school groups in the spring. Yes, we went to school, too. My older brother, Eric, my younger sister, Renee, and I would hop on the school bus in the morning. Mom would come get us out of school and back to the farm zoo by 10 am. Dad would hop in the Snake Den and give an educational lecture to each class. Then one of us kids or Mom would pick up the group and take them around to the different cages talking about the animals.

Johnny was the 2nd stop on the tour. I do have the tendency to be gabby (my childhood nickname was "Chatterbox"). Dad would often give me "the signal" to wind it up and move on, supervising the movement of groups in the zoo from the Snake Den. On one particular tour, I was talking about Johnny while feeding him chocolate from my shirt pocket. One half a square of Hershey bar was all he was allowed, as we had a lot of school groups that day. Dad would not allow us to screw up Johnny's nutritious diet with too much chocolate.

Johnny wanted more. I moved away from him, and he came to me. Johnny then proceeded to climb up my leg, with his tree climbing claws, smelling the chocolate in my shirt pocket. The school kids all thought it was part of the tour. I was in trouble. Johnny's tree climbing claws were hooked into my jeans. The only way to move Johnny was to carry him by his front claws pretending your fingers were tree branches. If you tried to pick him up bodily, you risked getting stuck with quills. Thinking on my feet, while still gabbing up a nervous storm, I took the chocolate from my shirt and enticed him back down my leg. This took a lot of time and by then Dad was whistling **and** signaling. I learned after that to use peppermint lifesavers and leave them outside the cage. When I went back to school that afternoon my knees were still shaking.

Chapter 6
Isabel the Mouflon Sheep

As I grew older, I was given more responsibility with the animals. The first infant I was totally in charge of, under my parents and Gramma's supervision, was Isabel. Isabel was a Mouflon sheep, a type of wild mountain sheep native to the Middle East. The tricky part with Mouflon sheep is the time of year when the babies are born. Typically the young are born during the last big snowstorm in March. In New Jersey, the last snowstorm is usually followed by the break of spring weather. I was 15 years old when the sheep were born early that spring.

Isabel was born in the middle of a big snowstorm in March, followed by an unusual cold snap. When Dad brought her in to me, she had frostbitten ears, tail, and hooves. We warmed her up and got her to drink. Gramma always told me to name sick babies after people in the Bible because it gave them a guardian angel. I flipped open the family bible, pointed my finger and closed my eyes... Isabel. Isabel was about one foot high and one foot long, all of her height in her legs. Soft tan fur covered her body; her eyes were black, with black rims, circled in white. Her belly was a lighter cream color. She looked quite pitiful with her shriveled ears and tail. But, like Dad and his ugly baby buzzard, I loved her.

My little brother, Parker, had been born in December of the previous year. He was born with black hair and sideburns, and looked like Elvis. Parker was six months old when we brought the lamb into the house. Parker loved to sit and watch the lamb walk by, grab her and give her "loves." We had to keep an eye on both of them, because Parker would often "love" a bit too firmly for the infant lamb. So now I had two favorite pets.

Isabel was put on a bottle schedule and lived in a box by my bed at night. My sister, Renee, and I shared a bedroom. Isabel would roam around our room, and soon began jumping on beds and desks and dressers and anything else with some height to it. I was so proud of her abilities. She was, after all, a mountain sheep, doing what came naturally. She loved to play with hair rollers (ladies—remember them?) and run after anything that moved.

Isabel had imprinted on me and accepted me as her mom. She followed me all over the place; "Mary had a little lamb..." In the morning I would deliver her to my Gramma at the general store. Gramma lamb-sat while I was at school. After school I'd report in and pick up Isabel, who'd follow me as I did my chores. At that time we lived in an apartment over a five bay garage. Stairs were no problem for Isabel, who would follow me up and down stairs easily.

One day, I had to deliver a message to Dad, who was working in a room over the slaughterhouse. It had an open stairway. Isabel followed me up the stairs, helped to deliver the message, and followed me back to the stairs. I started down the first step. Isabel decided to take a short cut and jumped off the top stair to the cement floor below. I became hysterical as I looked down and saw her motionless and spread-eagle amongst the machinery stored below. I ran back to my dad, crying.

Dad put his arm around me as I cried out my story. Then he said words I'll always remember: "Let's go see if she's breathing, where there's breath, there's hope." We walked out to the stairs and peered down to the dusty concrete floor below. She was not there. We went down the stairs and looked around the discarded machinery kept there. Dad said, "Call her, she'll come to your voice." I called to her, and Isabel came running, wiggling her wimpy tail behind her! Dad laughed and I cried tears of joy. Isabel was fine, a leaping wonder, and a true mountain sheep. After the stair-jumping incident, Isabel was confined inside the zoo grounds. She roamed free to pester every visitor that came with a baby for what she was truly after, a baby bottle.

Later that fall a feral dog jumped the zoo's perimeter fence and ran Isabel into the ground. Dad took her to the Vet. She came home with a lot of stitches in her neck and hindquarter. I was home from school, but knew nothing about it. My brother, Eric, came to get me when they drove in after the Vet had stitched Isabel's neck, shoulders and hind leg. I cried all the time we were setting her up in a special place in the garage beneath our apartment. Dad hugged me bear hug style. I knew things were bad when I looked up at him through my teary teenage eyes, and his wise farmer's eyes were crying too... Hours later Isabel passed away as I cradled her stitched, zippered head in my arms. I was inconsolable. The next day Dad raised all the perimeter fences around the zoo to eight foot high.

During those years Dad was the Sussex County Agricultural Representative to the New Jersey Fish and Game Council. My dad, Fred Space, helped to establish the criteria for keeping both wild and exotic animals throughout the state of New Jersey. Former Governor Hughes appointed him to the Fish, Game and Wildlife Council. One of the criteria for having any exotic animals today in the state of New Jersey is to have eight-foot high perimeter fences surrounding the zoo to protect the inhabitants from intruders as well as the outside world from escaped zoo animals.

Dad and my two favorite pets:
Baby Parker and Isabel, the Mouflon sheep.

Chapter 7
Raccoons and Kids

As I have mentioned before, on spring days when the zoo was busy with school trips we Space kids were brought home from school to give guided tours. I loved to do this—I inherited my dad and grandfather's ham bone. The fourth stop on the guided tour was raccoons. I had my dad's version of the "talk" down pat. Part of the speech on raccoons was to explain the concept of nocturnal. I would tell the teachers and children the definition. Several times I would say "NOC-turn-al" very distinctly to teach the kids a new word. One day I had a teacher who really worked with me. She would ask the children questions about the previous animal as we moved from enclosure to enclosure. Moving away from the raccoons she asked, "OK Children, who could tell me something special about raccoons?"

The cutest little girl answered right away, "They are NOT-turt-les!"

Ya know what? She was right.

Chapter 8
White Tail Deer Fawns

Gramma always raised the white tail fawns "found abandoned" by neighbors and strangers. Back then we used glass Coke bottles, long black rubber lamb's nipples and a special homemade formula. After a while the fawns would come running to the sound of the clinking glass bottles. For many years the fawns were allowed to roam the zoo grounds for the summer. They were people friendly and a big hit with the visitors.

I mention this as an educational message to my readers. Fawns are never abandoned if they are healthy. Mother Nature has designed the spotted coat of the fawn to camouflage it in the wild. In dappled sunlit forests or grassy fields the fawns disappear. Fawns also have no smell. These two facts keep the fawn safe from predators when it is too young to outrun its enemies. The doe (mother deer) returns to feed the fawn whenever necessary. If a young fawn is approached, it will lay down, flat and still, to hide. Mom is nearby, waiting for the danger to pass, confident in the camouflage nature has provided. Humans make the mistake of thinking the fawn is abandoned. As soon as the danger perceived by the doe is gone, (yes, this could be you!) she will come back to feed the fawn. It is not unusual for a doe to leave a fawn for hours to feed or drink for herself. Left alone the fawn waits as instructed, quiet and still, until Mom returns to feed the fawn. When the fawn is quick enough to run with its mother, it does. Our enjoyment of outdoor recreation should include looking, but not touching. Mother Nature takes care of its own.

Raising fawns with my Gramma is a treasured memory for me. I loved to see the fawns come running to the musical clinking of our glass bottles and the sound of our voices calling, "Come on, babies." They would charge down the hill with a frolicking bounce, white tails waving. By the way, fawns love to dance in the rain.

Chapter 9
Vacations and Holidays

Daily farm life is hard work. The three of us kids worked right alongside our parents and grandparents. When I was very young I remember being so proud the day I grew tall enough to reach the admissions cash register standing on a metal milk crate. Admission was 35 cents an adult. Eric worked with my dad with the Holstein cows and my grandfather in the mink ranch. Renee and I helped out in the restaurant bussing tables and taking orders until our change counting skills were good enough to run a register. That was after our daily chores were done outside. Kid type fun (baseball, tag etc.) was after the zoo closed and all the workmen went home at six p.m. What is called family time today was a non-existent concept then, at least on the farm. Every minute was spent with the family, working.

We did go on "vacations," but not the typical family vacations. The zoo was open all spring, summer and most of the fall. The cows needed to be milked twice daily, and the hay season was hard, hot, backbreaking labor. The mink ranch had breeding season, inoculations (every mink received an inoculation for distemper, enteritis, and botulism), blood testing (for Aleutian disease), kit season (time when the babies were born), grading (the time we checked each and every mink for quality) and pelting season. The time for family vacations was squeezed in between. Usually there was one vacation in the fall and one in the early spring on alternating years.

Our fall trips were tied into a hunting season out west, since Dad had friends all over the country. He would drive the family station wagon all the way, three kids in the back seat squabbling, as kids do. We visited all the national monuments along the way. Dad and Eric hunted at our destinations; we ladies kept ourselves busy in the travel trailer.

Five memorable situations stand out in my mind. When we visited Death Valley it poured rain, the cacti bloomed in time elapsed speed. I shot my first rabbit in Nebraska, and we ate him for breakfast. Outside Pittsburgh PA is a field of boards painted different colors (the Pittsburgh Paint Company's testing field). Mount Rushmore is larger in life than can be imagined. Finally, in Colorado I was so cold I slept all night with a woolen cap on my head. I was in every state except Hawaii and Alaska before I was fifteen, and that is all I remember.

We would stop at zoos and ranches picking up animals to bring home on the way back. The animals would ride in the trailer in the daytime and be switched to the back of the station wagon at night.

We fished in Canada, the travel trailer pulled into remote wilderness areas. We were up before dawn and out on the lakes to catch fish. One year we caught a young porcupine and smuggled him over the border in the trailer. After we crossed the border we stopped down the road to check on the porcupine and he was gone! We found him later; he had chewed out of his box and was roaming in the back section of the trailer. Good thing the border guard did not check under the bunks!

I had always wanted to see a moose in the wild. My brothers, Eric and Parker, went on multiple hunting trips with Dad in Canada and Alaska and had seen them along with a cornucopia of other wildlife. My dad had shot one in Alaska, but I never got to go. I had heard stories of how huge and cumbersome they were. It was one animal I had in my "life book" of things to see or do.

The spring vacations were a different matter. The family would pack up and head for Florida. On the way we would stop at every zoo along the way and visit Dad's colleagues. Seems Dad knew everybody in the animal business. Snake men, alligator men, hoof stock and lion people—all can be a colorful lot. These were very interesting folks with a million animal stories to tell and scars to prove them.

Once we arrived in Florida, the fun would begin. We would cruise the old canals and waterways looking for snakes. Snakes like to sun bathe, then slither into the water at the first sign of danger. Water snakes, cottonmouth moccasins, black snakes etc. would sun themselves on the banks near the water. Mom would drive the station wagon, Renee and I in the back seat and Dad and Eric standing on the open tailgate. When Dad or Eric spotted a snake they would pound on the roof of the car, signaling Mom to stop. Mom would slam on the brakes. The guys would slither down into the canal, catch the sunbathing snake and bring it back to us. Renee and I had to hold the snake bags open for Dad or Eric to drop the snake into. Next, Dad would swing the bag to twist the top and knot it. The snake bag would then be gently placed in the back of the station wagon, we'd all hop in and we would drive on looking for more snakes, repeating the process. I am sure we looked like a bazaar version of a Chinese fire drill to the casual passerby. At night the snakes would be kept in the travel trailer on the floor, if it was chilly outside. There were no nighttime bathroom visits!

Mom always asked if we could visit the ocean and see the seashore. We stopped once for twenty minutes. That was all I saw of the beach in Florida until I was eighteen.

Renee and I were 4-H'ers. Mom and Dad wanted us to be well rounded, so we learned how to sew (Mom was an accomplished seamstress) and make cookies. Back then girls did not participate in other 4-H activities. The best kid times I had were at 4-H camp, located twenty minutes and an entire world away from the working zoo and farm life. I was a Girl Scout also and once went to that camp. But I loved 4-H camp and went one week out of the summer from the time I was nine and then became a counselor until I was twenty. Surrounded by other animal oriented kids, I felt accepted. At 4-H camp I was not so much the weird kid who lived at a zoo.

Holidays were always spent at the farm. The workmen had holidays off so we would do the animal feeding and watering, but no extra projects. Animals (with the exception of snakes and hibernators) eat every day, they don't care if it is a human holiday or not. Fourth of July was during the open season, so visitors, workmen and the family were there. Labor Day, the crowds were here so we worked even harder.

Thanksgiving is during the hunting season. The menfolk would have breakfast before dawn at Gramma's house, go hunting in the mountains, come back to milk the cows, feed and water the animals in the zoo, feed and water the mink. Then they would come in at one p.m. for the big turkey dinner Mom was famous for. Like other families, Thanksgiving dinner took days to prepare and was wolfed down in twenty minutes. Then the men would milk the cows again, and be out hunting till dark. We ladies got to clean up and visit with the other women of our family.

Christmas for a farm family is different. Christmas morning my sister and I would wake up and rush to the Christmas tree. We would ogle the wrapped gifts waiting for Dad and Eric to come in from milking the cows. We were allowed to open our Christmas Stockings, but everything else had to wait. I know this sounds like a cruel torture, but I now know that Dad wanted to enjoy his kids' delight in opening the presents he worked all year to provide. After breakfast at my Aunt Loretta's, we'd go to Christmas Dinner at Gramma's. In between and over and around, the animals were all fed and watered, and the cows were milked again.

I loved New Years Eve. My family did not drink, but we kids always had a great time. Every farm had its own dump in those days. Ours was out in the backfields. When weather conditions were safe, Dad would set the dump on fire to burn the paper products and help the bio-degrading process. Every dump has rats. They were the same pests they are today. Rats carried disease and we could not chance them transmitting those diseases to the mink ranch or zoo animals. New Years Eve was one of the nights we set the dump on fire. Since we are a hunting family, Dad was adamant we were trained in gun handling and safety. My family is full of hunters and it was necessary we learned gun safety as youngsters. On New Years Eve my cousins, siblings, and I would go to the dump while it was on fire and shoot the rats running from the blazing dump. We became very good shots. Farm dumps and burning trash is now illegal. Our early family traditions were certainly different.

On New Years Day we traditionally went to my Gramma Ollie's (my mother's mother) in the "city" of East Orange. The cows were milked and the animals taken care of first. The cows were milked again when we got home. You can imagine how long those days were!

Washington and Lincoln's Birthdays were days off from school. Those days were spent in the mink yard. January is breeding season, so the male mink were carried to the female's pen, then back to their home when the breeding was complete. This often carried over into February.

Easter was a day for us kids to go to church. Of course, along with the other farm families, we milked the cows and fed the zoo animals first.

Memorial Day the zoo was open again, and we worked. Holidays were more hassles than fun, but a break in the daily routine, nonetheless.

Chapter 10
Salamander Hunts

At the drop of a hat and more like when our chores were done (before anyone found us and gave us more work to do) the cousins, Renee, and I would head out to the woods and go salamander hunting. We would put on clothes to get wet and dirty in and go sloshing through streams and fields. Looking under rocks, rotten logs, through mosses and in the streams themselves, we collected, counted and identified many different types of salamanders. Salamanders are amphibians. Amphibians have moist skin, and spend part of their life cycle in the water. They have the body shape of a lizard. Salamanders come in all sizes. Some we caught were one inch long, and some were a good eight inches! In our part of New Jersey, there are no poisonous salamanders but we always had to watch out for snakes. On one hunt we caught 117 salamanders! A record catch. We would bring them back to the zoo, set them up in aquariums, identify them, watch them for a few days, and then release them. I still love to salamander hunt and have taken groups of friends' kids to hunt and observe salamanders in the back woods and streams. We always catch and release, otherwise we could not go out and find them again.

Chapter 11
Boris the Otter

When I was a sophomore in high school Dad bought a litter of Asian otters. There are over twenty species of otters, with the Asian otters being one of the smaller species. There were five otters in the litter. About five months old, the litter made a great exhibit at the zoo. The otter enclosure was twenty by twenty feet, with a large pool on one side. The rock den was to the left of the pool and had a cement slide running from the top of the den to the pool. The litter would splash and slide all day long. Otters are nature's most playful creatures. Otters in the wild have been observed sliding down muddy banks into lakes and rivers. In winter they slide down the snowy banks into the icy water. Their thick fur keeps water off their skin. Dad had constructed the enclosure, imitating nature.

All went well, the otters loved the enclosure and entertained the visitors to the zoo daily. We do close the zoo in the winter, though. To winterize the zoo, the fronts of dens are enclosed to keep the cold air out. The dens are stuffed with hay. It snows heavy in the northern part of New Jersey and it gets cold. The otters still came out to play and swim every day.

A litter is considered to be all the siblings born at the same time. Every litter has a runt. A runt is the smallest member of the litter. The runt of the litter does not usually survive. Nature is tough that way. This litter's runt was carefully watched to make sure it had enough to eat and that the other sibling otters did not pick on her. Dad brought the runt in to me one morning. The night before had been cold and the runt had been kicked out of the den. Otters have four webbed feet (unlike beavers that only have two). This runt had four frostbitten feet, a frostbitten tail and ears. The extremities are the first to get frostbitten. Dad helped me set the little otter up in a small cage inside where it was warmer. We stuffed the inside of the cage with hay. Dad held her while I gave her a shot of penicillin, and a shot for botulism, just in case. We kept a supply of medications on hand so we only had to check with a vet for occasions like this.

An otter that has not been hand raised is still a wild animal. The runt wiggled and squirmed while getting her shots, although Dad held it tight. It tried to bite, of course, so we finished up as quick as possible and put the pitiful otter with ballooned feet in the cage with the soft hay. In all the hurry, we did not check the sex of the otter. I named it Boris, a goofy name for a goofy looking animal, goofy because of its ballooned, cartoon looking, frostbitten feet. I tended to Boris, with Dad and Gramma's help. Boris needed more shots. When its feet stopped swelling, the skin was swollen so much it cracked. The swelling finally went down. Boris was still not friendly. I used my Brother Parker's baby bathtub and gave Boris a place to soak in warm water. Some of the fur came off the paws and the bottom of Boris' tail. Otters' tails are usually completely covered in fur. Eventually, Boris accepted me, and allowed me to hold her. We found out she was a girl, but I kept the goofy name anyway.

Boris became a family pet since we were afraid the littermates would not take her back. Dad built her her own enclosure, which she was rarely in. She played and swam with us in our swimming pool. She was not happy alone, constantly chippering to all that would pass by her separated enclosure. When I went back to school in the fall, Dad gave her to The Turtleback Zoo in West Orange, New Jersey. She went to the educational department and traveled to schools to visit children. I was sad to see her go, but knew she would have a good life.

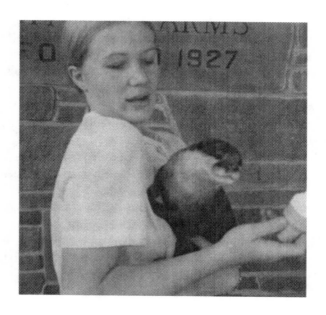

Boris Otter and I enjoy an ice cream cone.

Chapter 12
Bonnie and Benny, the Chimps

In the spring of 1971 two missionaries returned to the United States from Africa. The missionaries brought with them two six-month-old orphaned baby chimpanzees they had adopted. Their landlord would not let the missionaries keep the chimps. The chimps came to us in diapers and baby clothes. Dad built them a rumpus room (with bars) in the basement of our new house. Bonnie and Benny were adorable besides being very smart. They soon discarded their clothes and potty trained to a spot in the corner. I would take them out on walks around the farm. As they grew and became stronger I would walk them on a chain leash. Bonnie and Benny loved the gardens at Space Farms—loved to eat the flowers, that is. Gramma would get mad at anything that ate her flowers. Bonnie and Benny adopted me as their big sister and we would play and goof around daily.

Like other teenage girls my age, I would occasionally go out on dates. One (not so) gentleman caller took me to the movies. When we came home I wanted out of the car, so I asked him if he wanted to see the chimps. I was not very worldly and did not realize that that phrase was so similar to "want to see my etchings?" My date and I went to see the chimps. It was night so only a night-light was on. My dad's basement, where the rumpus room was, could in no way be considered romantic, but we are talking about a teenage boy now. I was talking to Bonnie and Benny close to the bars of their rumpus room. My date put his arms around me; I turned around and was facing him, with my back to the bars. Benny interpreted my date's arms around me as an aggressive act (or perhaps sensed my uneasiness with the situation). Benny's long hairy arms came out around my head and grabbed fists full of my date's hair, effectively pinning my date against me. I was sandwiched between my date and the bars. Benny was on the inside of the bars hooting and screaming, chimp talk for, "Don't you touch her, and I am upset about it!" Bonnie began screaming because Benny was screaming. My date was extremely surprised, to say the least, and upset! I wiggled around to yell at Benny to stop. Benny let go of my date's hair and scurried to his bed like a scolded little boy.

After I calmed the chastised chimps down, I turned around to see my date's car leaving the driveway, with him in it. I never heard from him again. My siblings still tease me; I am the "Only One Date" queen.

Bonnie and Benny grew up and were moved to the zoo. They were very smart; chimps have the intelligence of a five-year-old, scientists say. I think chimps are a very good judge of character. Often visitors would tease the chimps, so Bonnie and Benny developed their own special defense. The chimps would throw feces at the visitors that offended them. We put up a sign, stating just that. The chimps are long since gone, having lived many years at the zoo, but every now and then I'm tempted to put the sign up over the office door: "Caution, When irritated, these animals will throw feces!"

Parker, Benny the Chimp and I often went for walks in the zoo.

Chapter 13
Skunks

In the early sixties, a scientist for a perfume laboratory contacted my dad for the purchase of young skunks. The scientist was studying the musk gland of the skunk. We had skunks at the zoo, and bred them especially for her. We also caught pregnant wild skunks and kept them till they had their babies and released them afterward.

Each local farm had its dump (this was before the laws that prohibited open dump areas). Dad, Eric, Renee and I would go out in an old jeep with a large 55-gallon drum in the back. Armed with flashlights we would cruise the dumps looking for skunks. When we spotted the skunks we would shine the flashlight into their eyes to temporarily blind them in the darkness of the night. Skunks are excellent contortionist shots. Skunks can look you in the eye and spray you in the eye at the same time. Keeping the light in their eyes so they could not see us coming, we would slowly approach them and grab them by the tail. I'm going to let you in on the skunk secret that enabled us to pick them up without getting sprayed. The skunk's scent glands are located just inside the rectum. They are small grape sized balloons on the outside of the colon, inside the body proper. The nipple of the balloon extends through the colon wall just inside the rectum. If you keep the skunk off the ground, not letting their feet touch anything, they can not squeeze the muscles to contract the scent gland and spray you. I will admit that I did not pick up any skunks—I graciously left that job to my dad and brother. I was just a striped skunk spotter. The skunk scent is very potent; a teaspoon full of scent will permeate one square mile of airspace. This potency is what the scientists were studying at the time. A few years later the musk perfume line (for humans) was developed.

During that time skunks were also sold as pets. Dad developed an operation going through the rectum to remove the scent glands. My siblings and I often assisted with the operation. The local vets did not want to do this procedure because it smelled up their offices. We performed the operation outside behind a barn. And yes, we would end up very smelly. The old wives' tale of soaking in tomato juice holds true. If you soak the smelly item in tomato juice the odor goes away in a week. If you do nothing, the smell will go away in seven days. Take your choice!

One time Eric picked up a skunk, a heavy female, and the skunk's feet grabbed onto his pant leg. As he looked down the skunk got him right in the eyes. Eric was screaming and Dad was yelling, punctuated with laughter. Eric's eyes watered and he was moaning, "I'm gonna be blind for life!" Thankfully he was not. I knew why my classmates would not sit by me for days.

Chapter 14
Jingles the Jaguar

In 1971 President Richard Nixon signed the Endangered Species Act. Prior to this law any one could own an exotic animal. Space Farms is located about one and a half hours outside of New York City. In those days, it was commonplace to see people walking exotic cats on leashes down the sidewalk in New York City. Young exotic cats are adorable—until about four months old. Teenage exotic cats are a handful even for experienced trainers. Adult exotic cats (maturity is reached at three years) can never be trusted. It is instinctual for them to "go for" the back of the neck for the kill, or the hamstrings to cripple their prey. Big cats practice by playing and they do not differentiate humans from others of their species when raised by a human. Some exotic cats are more difficult than others are, each having its own personality. They are all very beautiful and very dangerous.

It was not uncommon for my family to find boxes of donations on the front steps of the office. Usually they were sick, injured or orphaned native wildlife. We would patch them up and if at all possible, release them in a remote area of the farm.

One box in particular was different. On the outside was Japanese writing. The only English words were "Jingle Bells." The box was whimpering. Dad took the box into the office. He had learned from years of experience not to open a container outside, as the frightened animal inside the box would always try to bolt. He looked inside.

When he called me in, I saw the most haunting, pitiful thing I have ever seen. There inside was a Jaguar cub, badly beaten, with a large scab on the front of its forehead. It was scared and trying to be tough, hissing and baring its teeth in a grand show of bravado. Correction: she was tough. Her eyes were glaring, daring me to touch her, to get close enough for her to rip me apart.

Jaguars used to roam the western United States, Central and South America. The jaguar was pushed south to the rain forest by human destruction of their habitat. Recently the jaguar has been seen in the southwestern United States once again. A stocky, spotted cat, they are often confused with leopards. The easiest way to tell the difference is by the spots or rosettes on the body. The back round fur is a rich golden tan, the spots are black. The spot of the jaguar is roundish with a spot in the middle; a leopard does not have the inner spot. There is also a black jaguar, this color phase is unusual but is carried genetically in the spotted jaguar. Short and chunky, up to two hundred and fifty pounds for the males, jaguars are known for their strength. But don't let their stature fool you, they are as quick and agile as their thinner African look-a-likes, the leopard. Jaguars are one of the big cats that actually enjoy water and are avid fishermen, flipping the fish onto the shore to complete the kill.

The small jaguar in the box was named Jingles. The box was the only clue as to where she came from, other than that we had no idea. Dad estimated her to be about four months old because she was the size of a full-grown beagle. Dad constructed a large wooden box, with wire windows on one entire side. The wooden box was set up inside the heated five bay garage, across from the zoo proper. We needed to keep the young cub in quarantine until we could determine the state of her health. We already had one Jaguar named Jackie and planned to put Jingles next to her as soon as Jingles recuperated from her wounds.

We don't know what happened to Jingles before she came to us. We soon discovered one thing: Jingles hated men. I was assigned to her. She was not friendly to me either, in the beginning. I was very careful around her, following my dad's specific directions:

#1 Always make sure the garage doors were closed before opening the wooden house's wire windowed door.

#2 Talk in a calm and soothing "baby" voice.

#3 Put food in the back left-hand corner quickly. (Toss it in there!)

#4 Quickly sneak out the water dish, clean, refill and return.

#5 Scrape out soiled hay slowly, with the door three quarters closed. I used a long handled hoe, keeping my hands as far away from the dangerous, young cat as possible.

#6 Keep talking in a soothing voice.

Jingles was a challenge. At first she would hiss and paw at me with claws extended. I must admit that in the beginning I did not get step 5 done to perfection on a daily basis. She would grab her meat and run to the back corner hissing and growling. She was trying to scare me. If she truly wanted to do me harm, I would have been shredded meat. I also knew not to push her buttons. I gave her the "space" she needed and worked around her. After three weeks of the same routine, she calmed down. Jingles would sit in the corner of her house while I did my work, singing and talking to her all the while. Her head had healed and the shabby scab was falling off.

Jingles' wooden house was raised off the concrete floor about two feet high. This was to help prevent the cold of the concrete floor of the heated garage from transferring to her. The wooden bottom of her house was covered in the hay I had to remove. In order to do this I squatted by the door with one knee on the backside of the door to close it fast if I had to. I had a couple of close calls in the beginning.

By the time Jingles had been with us six weeks, she became more responsive to me. She would "talk" to me, and come closer to me as I was cleaning, playing with the hoe. Jaguars vocalize to establish territory, the only North American cat to do so. Although Jingles had many sounds or cries, I learned to understand her wishes. It often takes many months to understand an animal's demeanor and vocalizations, so I was thrilled that Jingles responded to me so quickly.

One day, I was cleaning and she pushed hard on the door. She knocked me off balance, I had been squatting, and I ended up on my fanny on the floor. You can not imagine my shock when she came out of the house door and sat on my lap! I was petrified. Yet I knew I could not show my fear. Animals sense fear, they can smell it on your body. My fear slowly turned into fascination as I watched her rub her head on my arm. The new fur growing in, where the scab had fallen off, was itchy. I slowly raised my hand and rubbed Jingles' head. I petted her, then I rubbed her nubby head. Thinking all the while, *Dad is going to be mad if he has to take me to the hospital.* Heck, I was a young girl and did not want a scarred face! After ten minutes she decided she was hungry. Jingles jumped back into her house and started to eat. *Whew!* From then on, I would let Jingles out to run around the garage while I cleaned. All the men knew to keep away from her. Jingles would hiss, scream, growl and paw toward them if they came near her house. Jingles was my responsibility. I was 16 years old, a junior in high school. I enjoyed my time with Jingles, she was special to me and very affectionate, so I probably took a lot more time than was necessary with her.

My older brother, Eric, was very involved with the farming aspect of the 400-acre farm. He drove tractors and did a lot of the hard work farmers do, at an early age. It is an important job. We grew all the grain and hay for the hoof stock in the zoo. Tractors were kept in the five bay garage.

One day when I was "playing" with Jingles, Eric opened the side door of the garage and walked in. Jingles immediately bared her teeth, growled and ran toward him. Eric ran out the door. I had never seen my 18-year-old, six foot, two hundred fifty pound brother run so fast! He had just cause. Jingles ran to the door, growled, then returned to me, meowing, when she saw her "territory" was free of intruders.

Jingles graduated from her quarantine and was moved to the zoo, where the men took care of her, carefully, very carefully. Jingles still hated men. She enjoyed watching the visitors come by. The visitors stayed behind a guard fence five foot back from her enclosure. I visited her daily, rubbing and petting her from the outside of her enclosure. She had grown too big and strong to "play" with. Jingles was a beautiful cat, the scar on her forehead had disappeared, covered in swirling, black rosetted, tan fur.

I was a senior in high school and in class one day when I was called to the school office. My Mom was there to pick me up. I had no idea why. It turned out that Jingles had slipped out of her enclosure. While my Gramma's brother had been cleaning Jingles pushed the door and escaped. The men tried to approach her, Dad called to her, but she would have nothing to do with the men. Jingles was running around the zoo, inside the eight-foot-perimeter fence, chasing the free roaming chickens.

I called to Jingles in my "baby" voice. She trotted to me and followed me back to her enclosure. No big deal. I was driven back to school. After that Jingles' enclosure was scraped and hosed out from the outside. She lived a long life, moving to a natural turf enclosure and reared young. Today Space Farms Zoo has fourth generation Jaguars.

Chapter 15
Bear Cubs

Bear cubs were always being born at the zoo. Black bear cubs are born in mid to late January. They are about the size of a 12-oz soda bottle when born, hairless and eyes closed. At birth they are fetal in appearance. We would leave the cubs with the sow (female bear) for eight weeks, then take them away and put them on a bottle. This made the cubs more handleable and friendly to the zookeepers and vets. Most of our cubs went to other zoos around the country. Space Farms has had great success with breeding bears of all types and was responsible for successfully breeding the last surviving pair of Hokkaido bears. Hokkaido bears were bombed out of the wild in Japan after the Second World War. The last genetically pure pair came to the farm and produced young that were shipped to Uneo zoo in Japan.

I did not raise bear cubs; they were my grandmother's specialty. I babysat occasionally, but cubs adopt to one mom at a time, and they would not eat well for anyone else, no matter how close the imitation.

Dad often loaned out animals for talk shows in New York City. Sometimes the animals were rented; sometimes it was a barter deal for free publicity. I accompanied Dad and two bear cubs on one trip to a late night talk show hosted by a famous personality. This particular deal was barter for free publicity. We were working with another famous animal handler whose TV show was sponsored by an insurance company.

New York City is about one and half hours from the farm and driving in the city is not easy. The cubs were restless and uneasy in the car. After arriving and cleaning up the cubs in the green room, I was heating up the cubs' baby bottles, when the host of the show walked by and said, "Who's that?"

"That's the bears' mom," someone said . The host walked on; so much for meeting famous TV people!

The Handler took the cubs from my arms through a curtain; my hands were the only part of me that got on TV. The Handler talked about the cubs with the host, and gave Space Farms Zoo an incorrect, lousy reference. To say my Dad was angry is an understatement. After dashing through the rain to the parking garage, the steam was literally coming off the top of my Dad's crew cut. A vision I can see to this day. Needless to say, we never worked with that Handler again.

A 6-week-old grizzly bear cub is a bundle of cuddle,
but not when they are older.
Photo credit: Dr. Leonard Rue III

Chapter 16
Moses the African Atlas Lion

Space Farms was always boarding animals. Close to New York City, we often kept animals overnight for TV shows, or longer; the owners needed a licensed place to keep exotic or native wildlife. Think about it. If you cannot take a dog or cat to most hotels/motels, where can you keep a five-hundred-pound Atlas African Lion?

In 1971, a lady wanted to board a male Atlas African lion. Atlas lions are sub specie of the African lion. Atlas lions used to roam the northern parts of Africa, the Middle East and southern Europe. These are the lions that ate the Christians. The last Atlas lion in the wild was killed in 80 AD. The distinguishable characteristics of the Atlas lion are the long dark brown/black mane of the males, that extends all the way around their heads and down their chest to their loins, their larger size and their honey colored eyes. Dad checked the lion out for health and said we would board him. The lion was a circus lion, used to people, and did well at the zoo. There was only one problem. After a couple of months the lady never came back or paid her boarding bill. We never heard from her again. Space Farms had an Atlas African lion for keeps.

Dad arranged to get an old female African lion so the male would have some company. When we were assured the female was way past the age of reproduction (equal in age to an eighty-year-old woman), we took her in. Dad built a state of the art (of that time) enclosure with a large room-sized den, a wooden platform perch, a half roof for shade and a section in the sun. A retirement dream home for two circus lions used to living in circus wagons!

The retired couple got along well… very well. The cage was cleaned out daily. We did not go in; we used a twenty-foot-long T-bar to scrape and then hosed out the cage. As I've said before, big cats can not be trusted.

In July of 1974, Dad was cleaning the enclosure, scraping the large round poop out of the corner. Big cats are just like house cats that way—they always go in the same spot. He noticed one of the "poops" was moving. It was a tiny lion cub! What a surprise from an old lioness! Dad took the baby, and put it carefully back in the den. The old lioness promptly took the cub right back out to the corner and put it in the litter pile. She wanted nothing to do with it. Dad had told me what was going on so we kept an eye on the little one, hoping the lioness would change her mind. In the meanwhile the cub was crying and crawling around outside the den within the enclosure. A proper lioness is very protective of her young and would keep the cub(s) in the den. After a few hours we decided to take the cub out.

Dad enticed the adult lions into the den with a chunk of meat. The large metal door was slid into place, locking the lions inside the den. The adult lions were used to this procedure since we locked them in the den whenever necessary. Dad unlocked the side of the outside enclosure and I went in to get the cub. The cub had crawled toward the door of the den. When I picked up the cub, I could hear the lions breathing on the other side of the metal slide door. I got out of the enclosure as quickly as possible with my shaking knees. The den slide door was opened and the adult lions came strolling out, oblivious to my excitement.

My mom and dad had divorced, and Dad was dating Mira, a nice religious lady. When Mira heard the story of the lion cub, she exclaimed, "That's the story of Moses, a baby abandoned, and picked out of the moat by a princess!" The name stuck, Moses' name that is, I've never been the princess type. It's good luck to name animals after the Bible, Gramma always said so.

Moses was just two pounds. Small as African Lion cubs go. Note: Atlas Lion cubs range from one to four pounds at birth depending upon litter size. I put him on a baby formula. Moses drank two tablespoons every two hours, round the clock. He was also kept in a box with a heating pad by my bed at night. In the daytime he went to the zoo and stayed in a glass incubator, so visitors could see but not touch. Or breathe on him! There is only one thing worse than a protective mother lion, and that is an overprotective adoptive mother lion!

Moses grew fast. He developed a bladder infection at two weeks old, which he was treated for. I often gave him baths with baby shampoo. Lion cubs do not walk at birth, and their eyes are closed for the first week. Often one eye will open first, which is what happened with Moses. He looked like a pirate! By the time he was two weeks old, Moses would stagger around barely able to get his fat belly off the floor. Lion cubs progress rapidly. At four weeks he was running, jumping and stalking. We added ground meat to his diet when his carnassial teeth broke through the gums.

THE ZOOKEEPER'S DAUGHTER

Renee and I taught Moses doggie tricks: sit, stay, roll over, and open your mouth. We did not have to teach him the instinctual things a lion knows, stalking, biting, and running, jumping and growling. Moses never bit us to harm us, just to play like a puppy does with its mouth, or to let us know we were pestering him. Ok, so lion cubs don't like to wear human baby clothes!

There is no peace in a home with a lion cub. You can be sitting quietly reading the paper and surprise! You have been successfully stalked and jumped by a ten-pound lion cub. Moses caused havoc around the house. He would jump on the furniture, claw the curtains, and pounce on you just for fun. (Or practice?) Moses was also territorial. We had to be careful when visitors came to the house. I was nineteen and dating at the time. One date got as far as the front door. I opened the door and he saw Moses, heard Moses' growl, saw his bared teeth and split. Oh well, he must not have been my kind of guy. The Queen of "Only One Date" had struck out again.

Dad was usually very helpful with Moses, giving advice, and baby-sitting when I went out on dates at night. That summer a guy asked me out to the drive-in, alone. Previously, I had only been allowed to go out with a bunch of kids to the drive-in. I asked Dad to baby-sit Moses. Dad said he was busy and could not. I asked Gramma, she had a Grange meeting and could not. No baby-sitter. So Moses went to the drive-in with us. My Dad was very smart in retrospect. There is no monkey business at the drive-in with a protective lion in the back seat of the car.

Moses was sort of paper-trained. He would do his business on an open piece of newspaper. He was also growing fast. Sometimes he would have all four feet on the paper but his tail would still hang over the edge.... Soon a large section of our home was covered in newspaper. You don't punish a lion; you just put down more paper. My sister, Renee, and I liked to sew. One day Renee was laying out a paper pattern on some expensive material on the floor and the phone rang. Renee left her project unattended... Dad bought Renee new material.

Moses was only two and a half months old when I went back to college that fall. He was my baby, and I missed him terribly. "Born Free" was a movie about a lioness cub raised by a woman and rehabilitated to the wild. The movie was broadcast on TV in mid-September. I sat in my dorm room and cried my heart out. I called home and convinced my sister Renee to come visit and bring Moses. Moses rode six hours to my school in Rochester, NY, sitting on the front seat, a seat belt through his collar. I was so happy to see them. My dorm mates were quite fascinated; they had heard some of my animal stories, but I don't think they believed me until Moses came to visit.

Moses lived in a large wooden enclosure on my dad's front porch while the family worked. After work, Dad would put Moses out on a long clothesline type dog cable for a good run and playtime. Moses grew quickly and was thirty pounds by the time I came home for Thanksgiving. I will not say that Moses played rough, but he was an African lion, so let's just say he played "lionly." Moses would rub on your legs and purr, soaking up affection. The next minute he would try to bite your Achilles tendon. It was instinctual play, preparing him to hunt on his own. As long as you understood that, kept a continual eye on him and never turned your back, you were OK. Playtime was continued through the winter, but by the next spring Moses was the size of a German Shepard dog. One day Moses ripped my dad's coat off. It was time to move him to the zoo. Moses was moved to one of the larger enclosures at the zoo. He had an area to romp and play. A large natural stone den with a sliding door enabled us to lock him out of the den while we cleaned it from an outside door.

Atlas African lions love the snow. Moses would play outside in the snow and go into his den and rub himself dry on the soft hay. We had to clean the den on a regular basis, replacing the wet hay with clean, dry hay. When Moses was two years old, and a good-sized cat of about 300 pounds, the area was hit with a big snowstorm. Dad was out cleaning the animals and replacing hay. It was Moses' turn. Moses was used to the procedure, the den slide was slid into place, and Dad started to clean out the old hay. Due to the snow, Dad was working off the back end of a snowmobile with hay bales stacked up on the seat. Because of the buildup of snow in the track of the slide, the slide door did not close completely. Moses decided it was playtime. He hooked his paw under the heavy metal slide door and tossed it up, squirming into the den. Imagine my dad's surprise when Moses pushed his way past Dad onto the snowy ground beside him. Moses wanted affection! He rubbed on Dad's snowmobile suited legs and purred. When a 300-pound lion rubs on you, they really push you around. Moses rubbed and pushed and rubbed right between my dad's legs giving him a "horsey ride." My dad was strong, but could not lift a squirming 300-pound lion two feet up into the den. One of the workmen, named Joe, happened to come into the zoo at this time. Dad hollered to him, "Moses is out, go get Eric!" Joe disappeared quickly on his mission. Moses continued to rub on my dad and Dad figured out Moses wanted to get dried off on something. Dad quickly grabbed a bale of hay and fluffed it into the den. Dad patted the hay and said, "Come on, old boy, let's play." Moses put his massive head and two front feet into the den. Dad shoved Moses' back end with all his might and slammed the den's outside door. Dad was leaning against the den recuperating from exhaustion when Eric pulled up with the tractor and a come-a-long. Turns out Joe, who had a severe speech impediment complicated by missing front teeth, was very excited when he told Eric of the situation. All Eric could understand was that Dad was in the zoo and the "motor was out!" So Eric, not realizing the gravity of the situation, did not hurry to gather the equipment necessary to tow the snowmobile out of the zoo.

Moses lived many years at the zoo. We gave him a mate and moved him to the natural turf enclosure when his father passed away. Today I am raising Moses' great grandchildren, my great, great grandchildren. Moses was special, a bright light in my life.

Moses played with teeth and claws,
but was always gentle with me.
Photo credit: Dr. Leonard Rue III

Chapter 17
Night Capers

The zoo was a big part of my life, the animals my friends, the grounds my home. The best part of the day at the zoo is the night. The visitors were gone (not that I don't like visitors, they are our bread and butter). The day's work was done. The nocturnal animals were on the prowl, while other species safely slept. Night sounds from the zoo float on the dark breezes swirling to the stars. I am a night person. I am the offspring of generations of farmers who arose at daylight to start their backbreaking day, but I don't function well, mentally, before noon. That part of me must come from my mother's tainted city blood.

As children, my siblings, cousins, friends and I would often camp out in the zoo for a night or two. Dad would put up the large canvas tent he used for hunting trips in the wooded section of the zoo. We would stock up on goodies from Gramma's store and wait for dark. We never slept much, spending most of the night wandering the zoo with flashlights, looking at the different species in twinkling starlight. We watched with wondering eyes to see what the animals did after dark. Did you know that when chickens and roosters go to roost for the night, they sleep so soundly you can sneak up on them and pet them! In the middle of the night, the peacocks' calls sound like women screaming. The timbre of a lion's roar is so deep it reverberates in your bones. The coyotes' and wolves' howls are so lonesome that chills run down your spine. In the darkness of night, moon shadows of tree branches reach out to grasp the air from your lungs; you were never quite sure what was there. We were safe, of course, camping in the zoo, but in a child's mind...

In my wild and crazy teenage years, I would invite my dates for a late night stroll in the zoo. Often with babies to feed, we would take the time after the feeding to sit and listen to the night music nature provided. I had keys to the zoo.

Once, and only once, my girlfriend Minny and our dates stopped at the zoo for a midnight tour. The four of us enjoyed the night air, watching the animals and their night antics. Afterward, I brought the four of us into the main building for sodas. I knew the alarm system and motion detectors did not (then) cover the corner of the kitchen. We stayed in the back corner of the kitchen gabbing, drinking soda and having a great time as young people do. All of a sudden a spot light was shining toward us and my grandfather's angry deep voice boomed out, "Who the Hell is in there?"

    I tentatively stepped forward and peered into the darkness. Imagine my shock and apprehension when I saw my grandfather's shotgun trained on us from a secret peephole. He could see from his apartment into the main building's interior! Members of my family are excellent marksmen, and half heartedly joked about shooting first and asking questions later. A poster in the office window read, "Protected by Smith and Wesson." Another sign read, "Forget the dog, Beware of the Gun."

    I stepped toward the light that was calling me. I knew I was going to die, sooner or later. *Now by shotgun might be preferable as to what might come later. My a## is grass. Gramp is so hard of hearing how could he have heard us?* If I didn't step up, my friends also might get shot at. There was no doubt in my mind that Gramp would start shooting if I didn't say something that very moment. I stepped into the light, head bowed, knowing it was time to atone for my sins. I don't remember exactly what I said. My words faded into prayer after a cavalier attempt at bravery. "Hi, Gramp, it's me, Lori..." I winced, head bowed. The experience certainly put the kibosh on the evening. That gentleman never called me back either. The Queen of "Only One Date" had struck out, again.

    And what punishment did I receive? Well, if I had enough energy at the end of the workday to cavort out late at night, Gramp figured I was not working hard enough. I avoided Gramp for a while after that adventure.

Chapter 18
Geronimo and Little Breeze, the Cougars

After my junior year in college, I married and stayed in Rochester, New York for the summer. I missed all my animal friends and visited when I could. Majoring in biology was very interesting with learning the differences between the species from the inside out. I learned a lot from college, but not nearly as much as I had learned and am still learning from my family's lifelong experiences. Can you imagine my surprise when I turned the page in my Zoology textbook and saw a picture taken by Dr. Leonard Lee Rue III (a long time friend of the family), of my father's hand holding a timber rattlesnake milking it for venom? My senior year passed and Dad asked my husband and I to come home and work the farm. We moved into the little house next to the office and went to work.

My grandmother had aged and I took over the nursery department, helped her run the restaurant, and worked in the zoo. My family did not believe in special privileges for "The Boss' Kid" and I shoveled my fair share of manure. My grandfather, Ralph, ran the mink ranch and I worked there also. We inoculated the mink looking like mad scientists with needles and tubes running through our clothes to keep the serum warm in the winter. We conducted blood tests on each individual mink (25,000). We toted males to females during breeding season. We all skinned the dead mink during the pelting season. My work in the mink yard was done in the winter, which is the off season when the zoo was closed. But my true love was the zoo. The zoo at this time was about fifty acres, divided into three sections. I was responsible for cleaning the front third of the zoo, the older concrete and bar section. Every morning we'd all start about 7:30 washing and sterilizing every cage. Bears, raccoons, porcupines, fishers, wolverines, cougars, otters, beavers, woodchucks, and skunks (deodorized of course) greeted me every morning. Some were animal friends I had helped to raise.

THE ZOOKEEPER'S DAUGHTER

My routine had not changed from my younger years. I was up every morning to feed the babies in the spring and instead of going off to school I jumped into the other zoo work. Everyone has the impression that working at a zoo is all fun and games. This is not true. There is a lot of back-breaking dirty work, cleaning and scrubbing, litter and garbage pick up, lawn mowing, painting and general upkeep that goes on daily. I was not pampered as the "Zookeeper's Daughter." If anything, I had to keep up with the guys outside, and on top of that, help with the traditional "woman's work" inside the gift shop and restaurant. But gee... What's life without a challenge? It was the early seventies and I wanted to break out of the traditional female role at the family zoo. There was plenty of work to do and no one complained when I took on more.

We had always had great success breeding difficult species. Space Farms has a great zoological reputation, both nationally and internationally, with our breeding programs. We supply the animals with clean, spacious environments, great artesian well water and natural foods. Combine that with good, fresh country air in a quiet natural setting, great medical care and nature will take its course. The distribution of live stock between the zoos is necessary to ensure genetic diversity in captive populations. Most zoos prefer to trade young stock, as they adapt better to new environments and (this is the big point) it is much easier to move a small animal than a large one.

Puss Puss, our resident cougar, had raised many other cubs that had been distributed to other zoos across the nation. Puss Puss had been hand raised and donated to the zoo. For whatever reason Puss Puss decided not to raise her last two cubs. Geronimo and Little Breeze were very small as cougar cubs go, weighing about a pound each. Dad and I made the decision to bring them in when Puss Puss did not show her normal maternal behavior after their birth. The cubs were eight inches long, tan with black spots and born with eyes closed. Aside from being on the small side they appeared normal. Geronimo was a strong name with Native American heritage, and Little Breeze got her name by hissing without sound as a newborn. I put them on the same formula as Moses had been, and they did very well. Eyes opened at two weeks, walking at three weeks and stalking at four. I took them everywhere with me. They would toddle around the zoo following me as I did my chores, then settle down in the office for a nap (them, not me). They grew according to nature's plan and we did have them declawed at six weeks old. I do not like to declaw, as it is a surgical amputation of ten digits, the claws of a cat attach to the bone. It was a pitiful sight, two adorable kittens with gauze boots on all four feet. All went well though and it enabled us to handle the cubs safely as they grew.

Space Farms has always had free-range chickens and ducks roaming around the zoo, dozens of them. Every now and then a chicken or a duck would fly into a carnivore's enclosure and nature would take its course. Hopefully, this would not happen when the zoo was crowded with visitors. Dad always said, "Only the smart ones survive!" A twist on Darwin's Theory for sure! When Geronimo and Little Breeze were three months old, they were still following me around the zoo in the mornings as I did my chores. Geronimo spotted a chicken and with a gleam in his eye, stalked and pounced. I tried to get him off the chicken to no avail. The killer instinct had set in. I called to Dad and he came by to see what the problem was. Dad said it would be unfair to punish Geronimo or take his prize away from him, as Geronimo was only doing what his instincts told him to. You cannot make a lion a vegetarian! I was so proud of my babies growing up strong and "cougarly!" The decision was made to cage the cats in the daytime when visitors came to the zoo. Today a chicken, tomorrow...?

I still took them home every night—they slept in the bed with my husband and I. They used a litter box. (Have you ever tried to find a litter box big enough for a cougar cub?) They had been mildly rambunctious in the house. This lasted another month. The family adage, "When the curtains come down, the animals go out," came into play shortly after they were four months old.

The first night any of my babies stayed away from me was always tough, not on them but on me. The animals were always supplied with a cuddly blanket, a surrogate toy, a warm place to sleep in, and they were safe. I would worry all night that they missed me or were afraid. Just like any mom sending her kids off to college—you know they are prepared, but you worry anyway. I must admit the first "night out" for any of my babies still worries me and I usually check on them at least once in the dark with my trusty flashlight. Inevitably they are sleeping soundly and I feel like a fool for not trusting my head instead of listening to my heart. But after I check, I feel better and can finally get to sleep. (*I wonder if the campus police will understand the frazzled mom peering in my daughter's dorm room window with a flashlight when daughter Jackie goes off to college?*)

Geronimo and Little Breeze did just fine without me that night and all the nights thereafter. They grew to be beautiful beige cats, long and sleek, jumping from the log trees in their enclosure to the ground with ease. I would play with them daily. They always remembered me, even after I had left the farm.

THE ZOOKEEPER'S DAUGHTER

I could call them from a quarter mile away and they would cat call back to me. They lived long lives, but thankfully never produced any babies, as they were brother and sister.

Chapter 19
The Pet Shop Years

Living at the zoo was not easy. The workload was tough, dirty, and working with family is always a challenge. My brother Eric had established his place as a part owner with my dad and grandfather. My younger brother Parker was also to be given stock in the company. I seemed to be going nowhere. I loved the work, but the ten-hour days, six days a week, not including infant-animal-round-the-clock care, was taking a toll on my marriage. My husband's brother in-law had a position in a major company in Pittsburgh, PA and promised to help him get a job there. So with a lot of deliberation, after two years at the zoo, we moved to Pittsburgh. I landed a job in a flower shop and was able to expand on the botany I had learned in college.

Animals had always been the center of my life, so I applied to the Pittsburgh Zoo and found out I did not live within the city limits so I was disqualified. Their loss! I also applied at the local pet shop, and was hired immediately. I quit the flower shop and happily went to work in the Squirrel Hill section of Pittsburgh. Squirrel Hill is the Jewish section of Pittsburgh. I mention this only because it was quite a learning experience for me. Culture shock for a girl from the redneck countryside of New Jersey. I found the people there warm and friendly, if a little wary of me. On the third day of work I went to the local deli, me, a chunky blonde with braids wrapped around my head, and ordered a ham sandwich. Like I said, I learned a lot.

I amazed my coworkers at the pet shop when a for sale pet skunk got loose and hid out in the pet shop. I explained to my boss, Alan, that I could set a trap and catch the skunk overnight. This skunk had been out a week and knew the layout of the pet shop well. I used my country girl skills and set up an inverted flowerpot with an apple suspended from a string in the center. The pot was balanced on a Y-shaped stick and had a brick on top of it. Sure enough, the next morning the skunk was under the flowerpot. You can take a girl out of the country, but you can't take the country out of the girl!

I started in the pet shop in April of my twenty-second year and by

September of that fall I turned twenty-three. A position as manager became available and I was promoted to Manager of the Fox Chapel branch of the pet shop chain. I was so proud when I got that position. This was something I did on my own. The owner of the chain, Alan, was a wonderful mentor and patiently taught me the business. I loved the pet shop and it seemed my gift of gab enabled me to sell, using my animal knowledge to help my customers and their pets.

My marriage, on the other hand, was not going well at all. I decided to complete my college education. In my senior year of college I had taken two extra science courses and needed three credits in liberal arts to graduate. I had intended to complete them in New Jersey, but as I had mentioned, time at the family-run zoo was precious. I took my last credits at the University of Pittsburgh and received my Bachelors of Science from Nazareth College of Rochester in the mail in 1980. I divorced my husband and continued to work at the pet shop.

I told special friends about my family's farm and would come back to visit as time allowed. I visited all my animal babies that had grown up and marveled at their size and health. A few friends came home with me to see the farm. They were amazed; maybe they thought I was kidding about the lions, tigers and bears! *(OH MY!)*

The pet shop was special to me and I loved it. Alan helped me with the business aspects, but I was on my own most of the time. I got into breeding guinea pigs, and a local hospital would buy them to study their lungs. I found out the lungs of newborn guinea pigs were being tested with a medicine to help prevent Hyaline Membrane disease in premature human infants. I made a number of very dear friends at the pet shop, friends I still keep in touch with today. They would stop by to visit and buy items for their pets.

Silly things happened at the pet shop. The birds would get out and we (I had an assistant) would have to catch them. I became quite adept at catching flying finches and parakeets with my bare hands. I don't think that I was all that coordinated, I think the birds flew slowly because they were caged birds and lacked exercise. I also learned a few bird-catching tricks, wet birds cannot fly (with the exceptions of ducks and geese), and birds always fly up to escape.

We sold fish in the pet shop. I became quite the fisherman, with a net that is. We would put the fish in plastic bags and tie them in a knot. Occasionally a bag would leak and we would re-bag it. One day a young man and his fiancee came into the pet shop and bought some fish. The bag leaked all over her expensive designer jeans. She was hopping mad. He simply asked me to re-bag the fish. Doug didn't care about the jeans; he was concerned about the fish. Doug came in to buy a fifty-five-gallon fish tank and all the equipment. Doug became a regular customer and we became friends. He asked me out repeatedly and six months later, I invited him over for dinner. At his company party that year, his boss asked him where he had found me. His reply? "I picked her up in a pet shop!" His wise guy sense of humor always makes me laugh.

Doug and I were married in Beemerville Presbyterian Church, the church right next door to the zoo, in 1983. We bought a house in Harwick, a suburb of Pittsburgh and I continued to work in the pet shop. Doug was very helpful to me and taught me how to fix the electrical aspects of the fish tanks in the pet shop. He designed a compressor system to run air to all the fish tanks and would stay nights after his work, as a diesel engine protection systems analyst, to help with projects I thought up for the shop. His crackerjack sense of humor kept me laughing.

I continued doing projects for the family zoo. In my spare time I would paint signs and send them home with whichever relative was visiting me at the time. Dad would call me from time to time and we would discuss the problems at the farm.

Life in the pet shop was always interesting. One assistant of mine went on to study marine biology and became the director of the Marine Life Research Center of Juno Beach, Florida. Another went on to study veterinary medicine and raise champion Doberman Pinchers.

I was happy at the pet shop. Oh, I made a few mistakes, like the time I put a three-foot, black with yellow speckles, Florida King snake in the same tank as a four-foot yellow corn snake. I thought only the California King snakes were cannibalistic. The next morning when I opened the pet shop, I was surprised to see only one snake in the tank! The Florida King snake had devoured the larger yellow corn snake and had a belly distended like Christmas ribbon candy. Turns out both the Florida and California King snakes are cannibalistic. Alan was very understanding. Another day I was cleaning birds and an expensive one flew right out the open door! I religiously closed the door while working with birds after that.

The pet shop was located in a long strip plaza, next door to a jewelry store. I had a rather large snake get out. I warned the jewelry store about the possibility of it crawling through the walls and told them to call me if it showed up. It did. Scared the bejesus out of a wealthy lady who swore she would never shop there again. When I caught the snake, the clerk laughed and asked if we should x-ray the snake to see if I had trained it to swallow diamonds. They must have thought I was a much better animal trainer than I am. I was often called by the other merchants to catch animals that showed up in their shops, mostly mice. The animals were rarely ours.

In 1986 Doug and I decided to start a family. Alan and I agreed that I could work pregnant. My doctor instructed me on the precautions of infectious diseases and pregnancy. I was very careful and the pregnancy uneventful. Almost... the biggest problem was mobility. In my seventh month a hamster escaped. It was hiding under a large low shelf. I spotted the escaped rodent and grabbed a fish net, got down on all fours and still could not reach him. So I lay down on the floor, on my side because of my very pregnant belly, extended my arm with the fish net under the shelf and tried to catch him again. A loyal customer and friend came in the shop, saw me thrashing around on the floor and ran for help. I was so involved in the capture, with my back to the door; I had not seen her. My neighboring merchants came storming into the shop and were staring at me when I rolled over, pregnant whale style, with a hamster in the fish net! We all laughed about that for a long time.

Doug and I loved to wilderness camp. We had gone camping for years and continued after I was pregnant. We would pitch a tent for a week in the remote-water-access-only areas of the Pennsylvania State Park. I studied the flora and fauna, fished, swam and read books. In my seventh month of pregnancy, we camped in our tent. As most pregnant women have to do, I got up in the night to relieve myself. The tent we had was not tall enough for us to stand up in. One night I woke up and needed to go outside. Unbalanced, due to my pregnancy, I crawled on all fours, with a flashlight in one hand and toilet paper stuffed in my mouth, to the tent flap. I unzipped the tent flap and flung it back. Facing me dead ahead was a skunk, staring into the beam of the flashlight! *What to do, what to do!* I kept the skunk blinded by my flashlight, knowing if he could not see me, he could not spray me. I slowly lowered the tent flap, and zippered it shut. I then counted the seconds till daylight, hoping the nocturnal wandering skunk would go away. It was a long, long night.

My grandfather, Ralph Space, (founder of Space Farms), died that August. Doug and I went home for the funeral. All the politicians of the state of New Jersey, and all of my grandfather's eclectic friends were there. The state of New Jersey named a day after him. It was quite a tribute to an amazing, somewhat eccentric, yet visionary man.

I started my maternity leave the week before my daughter was due. While waiting for the big day at home, my neighbor's dog was hit by a car. My neighbor was distraught and called me to help. Once again I found myself on all fours, this time in the middle of a busy street. A cop drove by and immediately circled back, seeing a pregnant woman down in the road and a woman crying hysterically next to her. I explained the situation, stabilized the dog and sent them all to the vet.

I went home to wait for the big day.

Chapter 20
Jacquelyn Renee

Jackie was born on October 16, 1986. I was surprised that I, a self-appointed hippy style "earth mother" ended up needing a Caesarian. Jackie had her umbilical cord wrapped tightly around her neck and could not descend. Oh well. Jackie is healthy, perky, and the joy of our lives.

I returned to work at the pet shop ten days before Christmas and a dear friend, Pat McCarthy, babysat Jackie until the first of the year. Alan, Doug and I agreed to let me try to bring Jackie to work. I set up a nursery corner behind the counter. It was tough going, with an eight-week-old infant at work, but I could not give either up. Customers and friends stopped by to see us, make their purchases, and to cuddle Jackie.

Doug would drop us off at the pet shop every morning with all the baby equipment, and pick us up at night. On Thursdays when the pet shop was open late, Doug would stop by and take Jackie home with him. I'd come home later. Doug would tease me, that Jackie had only pickles for supper. He was very supportive of the working mother concept. Weekends were for family time, but Thursday night was special time for daddy and daughter.

Jackie loved the pet shop. What infant wouldn't love to sit and watch goldfish swimming back and forth or brightly colored finches twitter from branch to branch? She had her play pen (which is truly nothing more than a cage for a child) and her choice of squeaky dog toys to play with. I would count inventories out loud in a sing song voice to keep her company. (Instilling numbers in her head?) Customers that had become friends would take her for walks in her stroller as they shopped the plaza.

Jackie started to walk at ten months and by eighteen months was potty trained. Her special little potty was behind the desk. Customers and friends would help with positive reinforcement by clapping and cheering. I have found positive reinforcement works with all species!

Animals kept Jackie busy, she was taught, as I was, to be gentle to all creatures. She would play with the guinea pigs and rabbits while I cleaned their cages. Jackie would stack dog dishes for hours, like most kids play with blocks. All was working out well with her at the shop... almost.

One day, I lost her. I was in a panic and ran to the neighboring merchants for help. We scanned the plaza to no avail. I hysterically went back to the pet shop to call the police; afraid someone had kidnapped my blonde, green eyed baby girl. As I went behind the desk to call, I spotted her. She had curled up inside the rabbit cage with her favorite blanket and the rabbits. She was fast asleep. An apple doesn't fall far from the tree.

Another pet shop chain had opened up down the street, putting pressure on the pet shop I was running. Alan and I discussed it and I had a choice to make: put Jackie in day care or... Doug and I talked about it and Doug asked me what I wanted to do. There was no choice for me. After having left behind or sold so many animal babies, I could not give my only human child over to a full time sitter. I stayed home with my daughter for the next six years.

The next six years were a blessing for me. Having worked all my childhood at the farm, I finally got to be a child again, with a daughter as an excuse. We did all the things suburban families did, picnics, mud pies, bike rides, the blow up pool in the backyard, learning to swim at a friend's pool, church activities, and a few things other suburban families did not.

Doug, Jackie and I loved to wilderness camp and fish. We had a bass boat to take us into the water access only areas of the State Park. Jackie was a natural born fisherwoman. At three years old she could cast a fishing pole 30 feet out into the lake. She insisted on fishing with lures, as the worms were her friends. Jack would play with the worms until they died, when we would substitute a live one. I would pick worms out of her sleeping bag. As bedtime stories I would tell her all the stories my readers are now enjoying.

We would come back to visit the farm whenever possible. I continued to paint signs for the zoo and help however I could. I had always painted things for, and at, the zoo, going home on vacations and painting the snake den or cages.

At my first parent teacher conference, Jackie's kindergarten teacher told me Jackie had a "very vivid imagination," telling stories about the wild animals she had played with. I laughed and explained to the teacher that my family truly owned a zoo and she had indeed played with jaguars, bears and fawns.

While I was a stay-at-home mom, I discovered my painting skills, learned how to stencil, and became a crafter on the craft shows circuit. As a youngster who painted concrete animal enclosures, I never knew the pleasure creative painting could bring. My paintbrushes have become comforting, like an old friend that I often turn to when I am upset.

As every good mom does, I joined the PTA and helped with projects at school. We painted an entire playground on blacktop. Jackie's kindergarten teacher encouraged me to help design and build reading centers for the school. I enjoyed the creative process and built, painted and installed reading centers in a number of classrooms. My favorite one was a large cardboard pet shop (of course!).

Jackie enjoyed her days working at the pet shop.
She played with all the animals just like I did when I was little.

Chapter 21
Meanwhile, Back at the Farm

Meanwhile, back at the farm, things were not going well. My grandfather's death had saddled the zoo and museum with estate taxes. My two brothers were squabbling, the difference in their ages (17 years) and abilities causing constant friction. The animal rights people were always on the prowl, they did not like the mink ranch, hunters, or zoos in general. There was a lot of bad press; zoo attendance was up (rubbernecking?), then down. After my grandfather's death, Dad had refurbished the zoo, changing the outdated cement and bar cages to natural turf enclosures. It did not seem to matter that we were successfully breeding animals other zoos could not. The stigma of bad press has followed Space Farms for years. My Dad fought it and developed a reputation for being hot headed. He was defending his "baby," his life's work, the zoo.

Fur prices were down. Hired help was hard to come by, no one liked to shovel mink manure, or the backbreaking work the mink ranch entailed. The anti-fur people always amazed me. They would think nothing of wearing leather shoes, while touting their philosophy of not wearing fur. Where do they think leather comes from? The fur is just scraped off! Those who push nylon fake fur should realize nylon is a product made from oil, a non-renewable resource that takes millions of years for the earth to produce. Ranchers reproduce fur-bearing animals; mink in particular will reproduce at one year old. Coyotes will reproduce at two years old. I am totally against endangered species being used for fur. Ranch-raised or nuisance animals are a different story. They are just like the chicken or beef on the dinner table.

In its heyday, the Space Farms mink ranch kept 25,000 mink, individually caged, with scientifically designed nutrition, carefully structured genetic breeding programs and medical plans. My grandfather won many top breed awards for his mink ranch. Space Farms mink ranch was the largest mink ranch east of the Mississippi. The mink ranch was sold out six months after his death.

My older brother Eric started to concentrate his efforts at the zoo. My younger brother, Parker, and Eric did not get along. Eric decided to sell out his stock in the company to Dad and leave. The zoo needed managerial help.

In Pittsburgh, Jackie was seven years old and I was contemplating going back to work. Doug had been at his job for a little less than twenty years and was bored with it. The company he worked for had just gone Chapter 11. Dad came to Pittsburgh and made us an offer we could not refuse, including stock in the company if things worked out. I had always gotten along well with both my brothers. I asked Parker what he thought, privately, and explained what Doug and I could do for the family farm/zoo. He stated there was never enough time. I told him that was what we would come back to do, to make the time to fix up the zoo. The animals were always treated correctly, but the facilities had become in need of repair and an image spruce up.

I knew Doug and I could help. I was to take up where my father had left off with public relations, publicity, public speaking, raise the animal babies that needed human intervention (my Gramma was 90 years old), and use my painting abilities to give the zoo a woman's touch, "You know, spruce it up," Dad said. Doug was to use his electrical, mechanical and plumbing skills to help fix the multifarious things that go wrong on a working farm, and lend a hand with the muscle work. Doug also has a background in computer graphics, which would come in handy for signs, maps etc. at the zoo.

We moved back to the farm in the spring of 1995. The house we were to move into on the farm was still occupied by my brother Eric and his family. They were building a new house that would not be ready until October. We moved back into the house that I had grown up in, moved in with my dad and Mira. Our possessions were in storage, our days were spent at the zoo, and then we went to Dad's house to sleep. It was a rough six months.

Doug jumped into all aspects of his new job, enjoying the variety of work after 19 years of diesel engines. He enjoyed being outside most of the time, although the winters can be a challenge. His sly sense of humor and wit lightened the atmosphere at the farm. His abilities quickly won him the respect of his co-workers. I immediately started public speaking, taking animals to campgrounds, Rotaries, schools, travel shows, Boy Scouts, Girl Scouts and other tourism opportunities to promote Space Farms and raise attendance. In order to do that I needed animals that were easy to transport, friendly and safe. I took my animal babies on the road.

Grampa Ralph Space was known worldwide
for his mink ranching expertise.

Chapter 22
Belly Rub, the Coyote

I arrived at the farm two weeks later than Doug had. I stayed in Harwick until June 3rd so Jackie could finish out her school year. Jackie enjoyed the massive amount of family in New Jersey. There were only the three of us in Harwick, eight hours away from my family in Beemerville. Now that Jackie had more family around every day was like a party to her. She stayed at the zoo with me while I worked and helped out when and where she could. She had always wanted a puppy. Pop Pop (the name my dad's grandchildren call him) suggested a coyote puppy!

The momma coyote had had 6 puppies four weeks previously, and the runt of the litter seemed to be failing. Most mammals that have litters will lose one or two, usually the smallest ones. They simply are not strong enough to compete against their more aggressive littermates for momma's milk. Parker and Dad went into the coyote's den and brought out a small female puppy. I was happy she was a female because she would not mark territory in the house. Coyotes are canines and just like a male dog will lift his leg on the furniture, so would a male coyote.

Jackie wanted to name her new puppy Poochie. She was old enough to be weaned so we gave Poochie milk on top of her dog food. Then in the time-honored method, I would take her outside and stimulate her to move her bowels and urinate. After she did her business, I would reward her with kind words and a belly rub. Poochie grew fast and at five weeks could run faster than we could. Coyotes have been clocked at 35 miles per hour. But she loved her belly rubs. Poochie would dance around us but not come close enough to catch unless we promised "belly rubs." Poochie's name changed to Belly Rub, because that was the name she answered to.

Coyotes have exceptional hearing. Their large ears capture the smallest sounds. Also, like other canines, coyotes hear on different frequencies than humans. A coyote can hear a mouse squeak fifty yards away, or under four feet of snow, according to Dr. Leonard Lee Rue III in his book *Furbearing*

Animals of North America. Coyotes are shy and elusive by nature; most people here in the east never see one, even though the coyotes have quite a population in the area. Coyotes in the west hunt in the daytime, but the eastern coyote hunts at dawn and again at dusk to avoid the more active part of the human day. The eastern coyote has adapted to suburbia, often mistaken for a feral dog. The easiest way to tell the difference is by the tail; a wolf will run with its tail straight out, a coyote with its tail between its legs and a dog will always have a curl up to the end of its tail. The Spanish in the southwestern part of the US originally named coyotes. The Spanish also coined the phrase "Mi Coyote," for a coward who always ran away with his tail between his legs. The Spanish were wrong, the coyote is no coward, he simply avoids dangerous situations. That sounds pretty smart to me!

Belly Rub went with me on the publicity circuit. As a small puppy she seemed to enjoy the attention she garnered. She would hop out of her crate and into my arms. I would walk amongst the crowd of listeners telling them about her and coyotes in general. Belly Rub grew fast, and at ten weeks old was an arm full to carry around. I started to put her on a leash. At one of the campgrounds, I let her out of her transport crate, and hooked her leash on. One of the campers had a dog with him at the speech. The dog spotted Belly Rub and lunged toward her, barking. Belly Rub was scared and turned around and bit me in the fleshy part of my right hand between my thumb and forefinger. I quickly scooped up Belly Rub with my left hand, and held her cradled in my left arm. The show must go on, right? I started my info talk about coyotes calming Belly Rub with my right hand. Speaking to the audience I had not noticed the blood dripping profusely from the bite. A small boy in the front row said, "Lady, you're bleeding!" I gave the rest of my speech with my right hand in my rear jean pocket. I left Belly Rub home after that and have always requested that pet dogs be left at the campsite or home. People don't realize that, although their dog may be friendly to other humans or animals, my zoo animals have never seen domesticated dogs and are afraid of them.

Chapter 23
Kiwi the Coatimundi

Coatimundi are South American cousins to the raccoon. A full-grown coati will weigh about 15 pounds and is the size of a housecat. Coatis have a twelve-inch non-prehensile tail, long snout like an anteater, a narrow body, and a muscular nose. They have the same coloration as a raccoon. The masked beady eyes and long striped tail combine to give the coati a comical appearance. Coatis are excellent climbers and jumpers, using the long tail for balance. The diet of the coati in the wild is varied between fruits, grains, vegetables, insects and small animals. Before state and federal laws were passed in the early 80s the coati were sold for pets in the US. They can be litter trained just like a house cat. Space Farms had a breeding pair of Coatimundi.

The coati cannot take the cold weather of northern New Jersey. Every fall we take the pair across the street to one of the heated barns on the property. The male and female are separated to prevent breeding and delivery during the cold months. In late April or early May when we open the zoo, the coatis are placed back into the zoo proper. The enclosure is laced with limbs and lofts for climbing. Gestation is approximately 60 days. In July the babies are born. July gave us four babies in 1995. Newborn coatis are ugly. No other word can describe them. Hairless, black skin covers bodies the size of a spool of thread. Newborn coatis' eyes are sealed shut for two weeks. The four legs, long snout and naked skinny tail give them the appearance of a large wiggly spider. I don't care for spiders much.

Our mom coati had a great maternal record. It was a good thing she loved her babies since they were so ugly only a mother could love them. She kept her babies in a hay filled wooden box supplied for her. To prevent accidents caused by the male, mom and babies, nest and all, were transferred across the street to a warm barn. The mom had been hand reared by a human and was, therefore, used to the amount of contact and care she received. Dad knew she was getting near the end of her reproductive life. It was decided I should rear a young female by hand to prepare for the next generation of coatimundi living at the zoo. It is much easier to work with a zoo animal that is not afraid of humans. At four weeks old Kiwi was chosen for me to raise. She was a little small; her brothers were growing faster than she was. (Hogging the mother's milk?) Female coatis are smaller than the males as adults are. By this time Kiwi was the length of my hand, fully furred, masked eyes shining, and very mobile. Jackie named her Kiwi, not for the fruit but for the noise she made. She quickly adapted to the bottle and was soon also enjoying solid food. She was very vocal and chippered and called to us constantly with her keeeeeeweeeeep noises. Coati in the wild travel in troops using verbal communication and tail signals.

Kiwi was in constant motion. She traveled on my shoulders as I did my chores around the zoo. Kiwi's nose was a fascination for me. Her black, shiny nose sat like a cork on the end of her snout. Coati noses have extraordinary musculature enabling them to wiggle and twitch their noses in and out of every crevice, looking for insects and grubs to eat—and to explore any of my bodily cavities that intrigued her. Oh Yuck! As she rode around on my shoulders she regularly checked for bugs in my hair and ears. She knew I kept "goodies" for her in my pocket and wanted them constantly. I used these goodies to get her to go wherever I wanted her to go. Most of the time it was back to her "home" enclosure to rest. I'm not sure who needed the rest since Kiwi was full of energy and into everything, like a two-year-old on caffeine. Most of the time, I was the one who needed a break from the constant supervision of a zooming critter.

Kiwi was always near me that summer. She loved to jump from person to person. We played this game with family and friends but had to keep Kiwi caged when the general public was at the zoo. Alice, a former zookeeper, and Daisy, a grandmotherly woman, both worked at the admission desk. They kept an eye on Kiwi when I had to be out in the zoo proper. They babysat Kiwi and spoiled her with extra bottles and treats. Kiwi was quite a hit on the lecture circuit that summer. She would squirrel around my body as I spoke, looking for goodies. Children's eyes would light up with laughter watching her antics.

Every night Kiwi would go home to Dad's house with us (we were staying with him until our house was available). Kiwi would run rocket speed through the house, up and down the stairs, around the couches and then jump on all of us. Dad, Mira, Doug, Jackie and myself had our ears and hair checked nightly for bugs. By the way, Kiwi never found any. Kiwi certainly was a hoot, she kept us laughing every night.

By the end of the season it was time for Kiwi to be re-introduced to her mom and dad. She went into the coati enclosure in the zoo proper. The brothers had been placed at other zoos. She loved the branches and limbs to climb and jump on. Kiwi's parents accepted her antics, the same way human parents chuckle at the antics of a two-year-old. We all visited Kiwi daily, gave her treats and talked to her from across the zoo grounds to hear her call us back. It was hard for me, but I knew she was born to be a coatimundi and she would be happier with her troop.

Kiwi would constantly jump from person to person exploring all nooks and crannies, and she loved Jackie's hair.

Chapter 24
Henrietta the Chicken

Every year Space Farms would receive phone calls from people who had purchased an animal and for whatever reason could no longer keep it. A lot of Easter ducks and chickens found their way to the farm after they grew up and started to make bigger "messes." We made no promises, and did not coddle the fowl. The smart ones lived, the dumb ones would fly into the carnivore enclosures. Here was a twist on Darwin's theory—the survival of the smartest! All of our fowl have the equivalent of a Masters of Fowl, masters of the quick escape.

Henrietta was a small run-of-the-mill white chicken. She had been an Easter present to a mentally handicapped teenager. We took Henrietta in and released her in the zoo with the rest of the free-range chickens. A hundred ducks and chickens roam the zoo grounds to the delight of our visitors, many of whom have never been physically close to farm fowl.

Henrietta was special. The girl that had raised Henrietta kept her pet tame and cuddled her. Henrietta did not associate with the other farm fowl. Henrietta thought she was a person. Every time a door opened, Henrietta would zoom through the door to be inside the main building of the zoo. We would shoo her out time and time again. Anytime a visitor would sit down on the grass, Henrietta would jump up and settle on the visitor's lap.

Henrietta loved little girls, as she had imprinted on her special friend. Little girls visiting the zoo would often find themselves being followed by this small white chicken. Some children loved Henrietta, many were petrified and fearful of a chicken that was "chasing" them. Henrietta was undaunted, looking for her "mom." She settled down after a few weeks and became a favorite of the workforce at the zoo.

My brother Parker would feed the carnivores and omnivores at the zoo every day about three o'clock. Henrietta would hop on Parker's jitney (golf cart) and inspect the foodstuffs in the back. She would peck through the ingredients, eating whatever appealed to her. The animal food was kept in buckets and the old uneaten food was put in WWII survival cracker tins. My grandfather and father had purchased train cars full of WWII survival crackers shortly after WWII. We had sold them for years as animal food. Never wasting anything, as farmers can use anything, the tins themselves are still in use today. Henrietta would pick through the buckets and the tins on Parker's hour-long feeding tour. Even Parker, with his macho on his sleeve, would wait for Henrietta to join him to feed. Everyone loved Henrietta.

Space Farms Zoo and Museum has a fast food restaurant, complete with French fryer for, what else? French-Fries! Every so often the French fryer has to be cleaned out and the old grease is put out on the back porch of the kitchen to be taken to the garbage Dumpster across the street. The grease is often put in old WWII survival cracker tins.

I was occupied in the main building one day when Bill Raab, our resident farmer, came in chuckling. He relayed to me the cause of his chuckle. Seems Henrietta had jumped up on the side of the eightteen inch cracker tin full of warm grease to see what was inside, then lost her balance and fell in. Shortly after Bill had come out on the back porch to enjoy his 10 a.m. coffee and heard a chirp, but could not see its maker. Bill looked around and spotted Henrietta standing on tippy toe claws with just her beak above the warm grease in the tin! Bill scooped her out and Henrietta ran under a bush.

I rushed out the back door to find Henrietta. She was drenched in warm oil. Her down feathers were soaked, longer white feathers dripping, feet slipping and sliding over the grass. I picked her up and brought her into the restaurant kitchen, the only place on the zoo grounds with warm water. I put her in a bucket with warm water and dish soap. (Wouldn't that be a commercial!) I scrubbed and rinsed, and scrubbed and rinsed again. Wouldn't you know it, Henrietta loved it. She had been washed before! Chickens hate water, and cannot swim. Imagine my surprise when Henrietta stretched out her wing to be washed, and when that wing was completed, lifted her other wing! Henrietta had received tender loving care from her original very special mom. After thorough washing I put Henrietta under a heat lamp to dry in Dad's office. The aroma of a certain fast food restaurant permeated the room. Henrietta is the only chicken I know to survive French-frying! After that incident, Henrietta decided to hang out with the farm chickens, no more house chicken status for her! Henrietta was in many adventures after that, wandering around the zoo, chasing after little girls.

Chapter 25
Snakes on the Road

Snakes are amazing creatures. They have the ability to fascinate and repel an audience at the same time. Space Farms Zoo is famous for the Snake Den and members of the Space family climbing in to give snake lectures. The general public is so apprehensive about snakes, that fear is based on ignorance. In the family tradition started by my grandfather, carried on by my dad and my brothers, I regularly took snakes on the lecture circuit with me.

This section of New Jersey (and the northeastern United States) has only two poisonous snakes, the timber rattlesnake and the copperhead. My grandfather and dad used to take the poisonous snakes out to lectures. The laws in New Jersey have changed and we are no longer allowed to take poisonous snakes anywhere without a police escort and notification of the New Jersey Division of Fish and Wildlife. I take a few types of native non-poisonous constrictor snakes with me to campgrounds, Rotaries, schools, Girl Scouts and Boy Scout meetings at which I am speaking. Most of the time, things go well...

I was giving a lecture at a campground in the rain. This particular campground did not have a pavilion or meeting hall. I was a trooper when the rain started and kept on speaking. Thirty people surrounded me as I spoke on the milk snake I had wrapped around my wrist and hand. I like to take milk snakes because they are similar in appearance to the poisonous copperhead. The copperhead has two tones of copper on it, a lighter salmon color and a dark copper penny color. The milk snake is much more common and has rusty, gray and black coloration. The copperhead has no black on it at all. The milk snake is commonly killed because of this human ignorance. Laws in the state of New Jersey protect the rattlesnake and copperhead. I do stress at my lectures that snakes are not evil, just different than us, without arms and legs, and it is illegal to harm certain snakes and immoral to harm the rest of them just because of what they are.

I was feeling confident, I knew my snake facts. The audience was listening intently. I got to the part in my speech and said, "Snakes are not naturally aggressive and only bite when provoked." No sooner had I said the words than the rain stopped, the clouds parted, the sun glinted off my wedding band. The milk snake reared its head up and bit me in the soft fleshy connective tissue between my ring finger and my middle finger. Women in the audience screamed. I started to bleed, blood running down my hand with the snake still attached. Constrictor snakes' teeth tilt backward, so I had to pry the snakes mouth open, by applying pressure to the hinge of the jaw, while pushing the snake's mouth forward and up. I put the snake in the bag, tied the knot and washed my hand. Then I moved on to the next animal in my speech. You have days like that sometimes.

 A mom donated Charlie the king snake to us. Moms and relatives need to realize that the pet reptile you purchase for a pre-teen boy will still be alive when he goes off to college and/or gets married. Charlie is a great snake, friendly, active and a non-biter. King snakes are cannibalistic so Charlie was kept in a tank in the office of the main building of Space Farms. He had been handled by his previous owner and liked the warmth of my hand. I started to use him on the lecture circuit. He visited the Orange Central School and was petted, one finger, one hand, by the entire 3rd, 4th, and 5th grade, without incident.

 I was invited to speak at a Boy Scout Blue and Gold dinner. I had taken other animals and presented them first. Then it was Charlie's turn. I grabbed the snake bag/ pillowcase and started to untie the knot, starting my speech, "I'm now going to bring out a snake..." I looked into the bag and Charlie was not there. Gone, vanished, just plain old not there! I continued speaking. I gave half my snake lecture while flipping and handling the empty bag looking around the room for Charlie. Then I spotted him, coiled in the corner of the room, next to a table of grandmas. All eyes were on me, no one had noticed Charlie the King snake coiled in the corner of the room. I altered my speech: "Snakes are not naturally aggressive, there could be snakes around you and you would never know it." I then walked calmly over to the corner and picked up the sneaky snake. He had escaped out of a small hole in the seam of the snake bag. One of the grandmas at the table swooned. I have carefully checked the snake bags/pillowcases for holes ever since.

 Jenny, our secretary, is a real trooper about animals. I have plopped newborn lions, covered in birth fluids, on her lap to help me clean or tie off umbilical cords. Cubs have pooped meconium (the first black sticky poop) on her and she has helped wipe bottoms on many new baby animals. Jenny takes all this kind of stuff in stride, just another day at the office. But... Jenny hates snakes. We have tried to help her with this problem, to no avail.

Charlie, like most snakes, is an escape artist. I kept him in the office and he escaped. We were all on the lookout for him. Dad and I decided not to tell Jenny that Charlie had escaped. We searched high and low, on the pretense of cleaning the office. As we were "cleaning," we heard a ruckus in the gift shop next door to the office. I looked at Dad, Dad looked at me, and we headed out the door. Sure enough, scooting across the gift shop floor was Charlie. I scooped him up, much to the amazement of the fifty school kids shopping in the gift shop. Charlie was put back into his tank in the office. Jenny asked me to move the snake, and I moved him into the main building proper.

Charlie escaped again that winter. We looked high and low (again) and finally gave up, figuring Charlie had escaped to the outside through holes in the floor for plumbing etc. Charlie was gone for all of the winter. I thought he had died outside in the freezing winter weather. One day in June, Parker was walking through the building to get coffee before we opened for the day. He spotted Charlie retreating under the ice cream freezer. Parker caught Charlie and put him back in his tank. Charlie's tank lid is now reinforced with duct tape to keep it closed.

Chapter 26
Winking, Blinking and Nod, the Kestrels

There are many types of hawks in New Jersey. The smallest one is the sparrow hawk (common name) also known as the American Kestrel. It is a small bird about the size of your fist, not including legs, when full grown. Reddish brown in coloration, the males have significantly more gray in the breast and on the wings. In the spring of 1996, Space Farms received a phone call from a local demolition company. They were tearing down an old barn and had inadvertently destroyed a nest with three young sparrow hawks. Did we want them? I hopped in the truck and went to pick them up.

When I arrived on the scene, three puffballs about the size of a chicken egg with huge blinking eyes stared at me from a cardboard box. Mom hawk was nowhere to be seen, and there were feral barn cats roaming around. I knew if I left them, the cats would get them before the men cleared the barn and mom hawk came looking. I brought them back to the zoo.

Young sparrow hawks are ravenous. Carnivorous birds cannot eat straight meat. In the wild they would eat small rodents, skin, bone, intestines and all. We did not have a supply of mice but we did have venison. The local roads were scattered with road killed deer. I knew when I released the sparrow hawks they would have a good supply of food until they learned to hunt rodents. I fed them strips of venison dipped in crushed ferret food. The strips were about the diameter of an earthworm, 1/2 inch long. I gave them a vitamin water mixture by eyedropper every two hours with their feeding. I named them Winking, Blinking, and Nod. Nod was the smallest, and would immediately nod off after he ate. I took them with me on the lecture circuit.

The three little birds grew in their pinfeathers and started to fly at about five weeks old. At that age I was feeding them three times a day. I would open their large flight cage; each would fly to my finger, grab a tasty morsel and fly back to the branch in the cage to eat. I repeated this procedure the same way every time. Occasionally, one would take a flight around the zoo and come back for his food, never flying for more than a minute or two. Nod stayed put, hopping to my finger and back to his branch. Every family has its slow one.

At lectures the threesome behaved very well. I would keep one on my hand and feed it, then they would hop back into their transport cage. Except for one time. I was speaking at a campground with an open pavilion. Most campgrounds allow dogs on a leash with their owners. I had asked that no dogs be brought to the lecture, however... As Winking was receiving his meat a dog barked and Winking flew off into the trees. I was mortified. Yes, I was planning to release the three sparrow hawks, but I did not feel they were capable of surviving on their own just yet. I am an overprotective mom. This was unplanned. The audience waited patiently for me to continue, thinking this was all part of the show.

I mustered up and explained that today was release day for the hawks, declaring how proud I was of their progress. I was almost in tears. I passed it off as a proud mom, packed up and went back to the zoo. Later that night I went back to the campground with a fresh supply of meat strips with no results. The next day I did the same, explaining to the campers that I was just "checking" on my baby. No results. At the last feeding that night at the zoo, I noticed a small bird circling above the nursery. Too high to see, but I had high hopes. I put Blinking and Nod in the transport cage inside the flight cage and left the flight cage door open. I put strips of meat on the branches. In the morning the strips were gone. Winking was nowhere to be seen.

When it was time to let the two small full-grown hawks go, I simply left the door of the flight cage open. Blinking and Nod flew out at dusk and were back in the mornings for a couple of days. Nod was the last to go. Then they were on their own. I could watch them circling the zoo daily. I left strips of meat out for them, until, one day the strips were still there in the morning. We now have families of Kestrels living in the large locust tree across the street from the zoo. We keep losing chicks from the broods of our free ranging chickens at the zoo... I wonder why.

Nod the Kestrel was trained to come for food treats.

Chapter 27
Syrian Grizzly Bears

Dad had redesigned and built a new bear exhibit in 1985. This new enclosure was a bear's dream come true. An acre of prime New Jersey hillside was enclosed in ten-foot chain link. The top of the chain link has a four-foot overhang at a forty-five-degree angle to prevent climbers from escaping. To prevent tunneling out of the bottom of the enclosure, the chain link extends four feet under the ground. Trees, including an old apple tree (any bear's favorite), were wrapped in chain link to prevent the inhabitants from scratching the bark off the tree and killing the tree. A recycled two thousand gallon metal storage tank was inserted into the side of a manmade hill in the center of the enclosure for a den. The grass, trees and native shrubbery added to the natural enclosure. A large swimming pool completed the posh enclosure. Three strands of electric fence, powered by photoelectric cells, added more protection from escape. When Doug and I came home to work at the zoo in the spring of 1995, two Himalayan bears called that state of the art enclosure home.

In anticipation of a major move of one third of the zoo that fall, Dad decided to move the more active Syrian Grizzly bears into that enclosure. This is the first enclosure outside the main entrance, visible from inside the main building. The Himalayan bears were couch potatoes and did not put on a good "show," sleeping most of the day.

The Himalayan bears were fed daily in transport cages, until the bears felt comfortable in the transports. While both bears were eating one day, the slide on the transport was dropped. The transport cages, with the Himalayan bears still inside, were moved by backhoe to the new similar enclosure that Doug and Dad had constructed earlier the previous summer on the other side of the zoo. The Himalayan bears adjusted well to their new home, and as before, slept most of the time.

The Syrian Grizzlies were housed in a USDA approved cement bottom enclosure in the western section of the zoo. These bears were retired circus bears, and very interactive with the public. It was time to move the Syrian Grizzly bears and like families everywhere, the Spaces faced it with anticipation tempered with a touch of trepidation. We were moving not one but two six-hundred-pound bears from the concrete enclosure to their new natural habitat. Every move was choreographed, the plan discussed and reviewed extensively. Would the bears like the grass they had never felt before? Keeping in mind they were retired circus bears and perhaps were afraid of things they had not been exposed to (i.e. grass), would the bears refuse to come out of their transport cages? Or after being shook up on the backhoe ride across the zoo, would they try to climb over the top of the five foot tall transport cages, doubling back on the men stationed there, to head back to their former home on the west side of the zoo? Anything can happen when working with wild animals, especially wild animals that had been in contact with humans and, perhaps, lost their fear of man. We were not working with stuffed Teddy Bears. There was a loaded gun on the seat of one of the trucks. Our staff knew all the possibilities. We thought we were prepared.

 I watched from my observation point inside the glass doors of the main building. My job was simple (and safe)—plug in the electric fence on signal. Parker's wife Jill, seven months pregnant with her second child and with her ten-month-old son, Hunter, on her hip, watched the scenario with me. Dad, sitting on the backhoe, gave the signal to begin the process. My husband Doug and Keith (another keeper) were perched atop the five-foot-high steel transport cages, with pitch forks in hand, just in case the bears decided to double back and try to go over top of the transports. Doug and Keith opened the slides. The bears hesitated. We worried.

 The first huge golden bear took a tentative step onto the grass. Lifting his feet like a cat walking on a wet surface, he took a few steps. The second bear emerged, very curious. Slowly both brave bears meandered a few feet more into the enclosure. Fifteen feet was the magic number. The bears stood sniffing the leaves and grass, looking around like Alice in her Wonderland. Finally the massive bears made the necessary fifteen-foot mark. Dad started up the backhoe, put it in reverse. The grizzly bears were frightened by the noise of the backhoe and ran over the hill to the back of the enclosure. Doug and Keith jumped down from the transport cages. Dad pulled the transport cages out of the gateway. Parker hooked up the electric fence, and gave me the signal. I ran to the electric fence socket and plugged it in. We all breathed a sigh of relief. The men were patting themselves on the back for a job well done. It was then that Dad saw the chickens in the lower corner of the enclosure. Amongst the chickens was an ordinary little white hen...

Dad quickly unlatched the chain link gate, grabbed a pitchfork, stepped gingerly between the strands of the electric fence and started toward the chickens. From our vantage point Jill and I were the only people who could see the curious bears over the hill, Dad chasing the chickens in the front lower corner and Parker, Doug and Keith standing by the gate to the east.

I must admit I thought my Dad was nuts. He was risking his life for some silly chickens. Parker was thinking the same thing, I could tell by the way Parker whipped off his hat, smacking it on his knee. He was mouthing words I could not hear but I knew what they were.

Parker picked up a pitchfork and followed Dad into the enclosure. He also stepped gingerly over the electric fence. Parker was not concentrating on the electric fence, but on our father chasing the chickens, while watching for the grizzlies to come back over the hill. Parker wears his pants low on his hips. The electric fence zapped him with a thousand volts. Parker pirouetted six feet into the enclosure, twirled on his tippy toes, then shook himself a few times while stomping his feet. With the pitch fork still in his hands Parker looked like a dancing Satan. Jill asked me, "Why is Parker doing that?"

I snapped, "I hope you don't want any more children!" I was concerned for my brother, a thousand volts would stop a bear, what would it do to a man?

The action had distracted me, it all happened so quickly. I ran to unplug the electric fence, still wondering why Dad would risk his life for a few dumb chickens. I've mentioned our family's motto before, "Only the smart chickens survive." Then it dawned on me. Dad thought the little white hen was our pet chicken, Henrietta! I quickly got on the loud speaker system and told Dad that it was not Henrietta (which I repeated twice). The grizzlies were starting to move, curious, no doubt, about all the hollering and cursing going on. I informed my brother and Dad that the bears were moving toward them over the public address system. Parker convinced Dad to exit the enclosure, leaving the chickens behind.

Parker and Dad climbed back over the neutralized electric fence strands. The chain link gates were closed and padlocked. I plugged in the electric fence again. The guys put their equipment away, and went back to their work. Dad took the backhoe and the transport cages across the street to storage in one of the barns. Parker seemed fine, he prides himself on being tough, though he did walk a little taller for the rest of the day.

THE ZOOKEEPER'S DAUGHTER

The Syrian Grizzly bears adjusted well to their new home, frolicking in the green grass, rubbing against the chain link wrapped trees, and nibbling on the shrubbery. And the chickens? The smart ones flew out.

Chapter 28
Monkey Business

At the end of October the Space Farms Zoo and Museum closes for the season. Certain animals cannot take the cold and must be moved to warmer quarters. Every year the Snow Macaques, a cousin to the Rhesus monkey, were shooed into a transport cage. This was no easy task. The men would go in the enclosure with brooms, rakes and shovels. Banging on the enclosure walls, the monkeys would be shooed into the transport cage. Chemical immobilization is not an option when the enclosure is two stories high, a drugged animal may fall and injure itself.

In the summer of 1995 we received a new male Macaque to help expand the gene pool of our troupe. The female Rhesus and the two castrated male Macaque were easily intimidated. They ran into the transport cage, the transport was closed and locked. The box full of monkeys was moved across the street to their warm winter quarters in a barn. The new male Macaque was the leader of the troupe. He was lord and master of his harem and eunuchs. He was not to be insulted or intimidated. Macaques and monkeys are incredibly strong. They can lift their entire body weight on one finger. Male Macaques weigh about 40 pounds. These particular Macaques were rescued laboratory animals; the head male was particularly nasty.

Dad, Parker, Keith (another zookeeper), and my husband, Doug, were all in the enclosure with the nasty male Macaque rattling the walls with brooms, rakes and shovels. There was a baseball bat, just in case. I manned the outside gate of the enclosure, ready to let the men out in case of mishap. The Macaque moved around and around the circular enclosure, avoiding the men and the transport cage. He was angry at the intrusion into his territory and the removal of his harem. He jumped from rope to rope swinging down and threatened the men with bared teeth. A monkey bite is a nasty bite. Their long carnassial teeth rip and tear. Strong monkey hands gouge and pinch in fights. A Macaque is nothing to fool with.

The men, with the exception of Doug, had done this procedure many times. They knew all the tricks monkeys pull on their keepers. Doug was new to the animal keeping world and was still learning basic animal behavior, his expertise being in the mechanical world. I watched the action from my post by the gate, keeping a special eye on my honey. The men were shouting and cursing, part of the intimidation procedure. The Macaque was not going in the transport box. The Macaque swung down by Doug. Doug turned his head and looked the Macaque square in the eye. I saw what happened and shouted to Doug, "NO Direct eye contact, no direct eye contact! It's a threat." Doug averted his eyes as the Macaque jumped toward him, screeching and teeth barred for a fight. The other men stepped in and scared the Macaque away.

   I did not like the situation; this male was not backing down. The men came out to take a break and I suggested another method. I had been feeding the Macaques candy as a treat. A few M&Ms here and there would not spoil their diet. Dad and Doug set up the transport cage with a trip wire triggered when the Macaque reached for the brightly colored candy. We left the area and waited for the Macaque to calm down and take the bait. It took a day but the candy did the trick. The head honcho Macaque was transported across the street to the winter quarters where he happily joined the rest of his harem. We've used the bait into the box method since, with a lot less stress to humans and monkeys involved.

Chapter 29
Moving Day

After the zoo closed for the season in 1995, we needed to move one third of the zoo off of land that belonged to my brother, Eric. Eric held the deed on the old Schellenger farm, the part of the zoo that housed grizzly bears, llama, buffalo, long horned cattle, yak, prairie dogs, woodchucks, cougars and bobcats. You can imagine the effort it took to move all these animals if you've ever had to move your dog or cat to the vet. What is life without a challenge?

The grizzly bears were timid, but after feeding them for days in a transport cage, finally we dropped the slide and moved them by backhoe. They were sent to Al Rix's Big Bear Mountain in New York State,

The llamas were herded out of their enclosure and moved rodeo style using trucks, with panels of fencing as a shield. They did not want to go. An animal gets used to its territory, and considers it home. The llamas were pushed and prodded with the fencing through the back path of the zoo to their new home bordering on route 629, on the outside of the museum complex. Our human neighbors enjoy the new view. Woodchucks and prairie dogs were netted out and the cement walls of their enclosures were moved by backhoe to the eastern section of the zoo. The woodchucks and prairie dogs were hibernated in the barns for the winter, their enclosure to be reconstructed in the spring. The bobcats were carefully netted and moved to the barn. Their paneled enclosure was reconstructed on the grassy plain next to where the buffalo were scheduled to go.

The American Bison or Buffalo are herd animals. Their massive size makes them dangerous, not to mention their kicking hooves and hook shaped horns. I have always wanted to raise one from a calf, however there are few things more dangerous than a ton of friendly buffalo. Buffalo express affection by rubbing their large, hairy, horned heads on each other. Space Farms Zoo and Museum is marked on the national map of buffalo herds. Our first buffalo was named Bertrum. His mate was named Bertha. They came to the farm before I was born. Dad had expanded the herd through a diverse breeding program. Twelve buffalo were to move to the eastern section of the zoo. On the day designated for the buffalo move, we opened their gate. Our dozen buffalo moseyed out munching on the green grass along the path. We were prepared for trouble and were surprised at their calm demeanor. Dad decided we would monitor their progress, slowly and more importantly, quietly. Buffalo are known to stampede.

A buffalo stampede is a swarm of irrational, fearful tons of hoofed animals on the run. When the Native Americans ruled the continent, buffalo numbered in the millions. Before the Native Americans had horses, they would trap the buffalo by stampeding the massive beasts, guiding the buffalo to stampede irrationally over a cliff. The Native Americans would then 'finish off' the buffalo after they fell and were too injured to escape.

Early American settlers traveling west in wagon trains noted in diaries that they had to camp for days waiting for multitudes of buffalo to pass before being able to continue their way west. A stampede can take down any fence, especially the flimsy temporary fences we had constructed in the zoo to guide the buffalo to their new grassy paddock.

The buffalo moseyed quite nicely up one hill, down the path and over the next hill. We were following behind on foot, not wanting to start a truck engine and frighten the huge bovines into a true stampede. The move was going better than expected. As a group, Dad, Parker, Doug, Keith, Bill and I cleared the last hill long behind the moving herd. Some of us humans move slower than the rest, Dad, Bill and I brought up the rear. To our surprise the buffalo had had enough of quietly grazing, and had doubled back. They were in a hurry, not quite an unruly stampede, but a frolicking trot that was not to be interrupted. Calves in the herd kicked up their heels in delight. The herd of twelve massive animals wanted to go home to their old paddock. We humans ran for cover beside the fences, large rocks and picnic tables. I was truly a smart chicken, and hid behind a tree. When the herd got to their former pasture, they found the gate closed, slowed down and milled about aimlessly.

Dad and Bill were behind me on the forty-foot wide path. Dad slowly meandered behind the herd. Dad removed his trademark cowboy hat, swooped it in the air and started the buffalo trotting once again down the path. Tons of buffalo thundered the ground as they passed by me. This time when the buffalo were close to the gate of the new acre pasture Dad started to whoop and holler. Those of us who were closer to the tail end of the buffalo herd followed his lead. The buffalo swarmed into the new paddock and Keith (the fastest of our human herd) closed the gate. We all congratulated each other and laughed about our close calls as we gasped for breath. Dad figured that was enough animal moving for one day. Thinking back on the situation, I can't blame the buffalo. If I were being chased by a group of redneck farmers, a six foot three huge farmer, a mechanic and a plump, middle aged, retired housewife, I'd run to new pasture, too!

The last of the large hoofed stock to be moved to the new section of the zoo was the Yak. Not quite as large as a buffalo, the yak still has an impressive size and longer, larger horns. The yak is a black-brown hairy bovine from the Himalayan Mountains. Domesticated for pack animals by the natives of that area, they are also used for milk, meat and fur. Space Farms Zoo was home to three females and a male. The original male and female were about thirty years old, very old for yaks. The other females were offspring of the original two. The yaks had lived in their acre paddock all of their lives. I had no idea how stubborn yak could be. The yaks' placid demeanor in their home paddock was deceiving.

The gate was left open for the yaks to wander the same path as the buffalo had. After two days the yaks had not come out of their pasture. Parker had tried to entice them out with special treats of food along the path (like Hansel and Gretel's breadcrumbs) to no avail. It was fall and there was plenty of grass and hay in the paddock. Next we moved the water dish. Not one yak ventured out of the gate. Dad decided it was time for more drastic action and a rodeo roundup using trucks and a tractor was organized. Tranquilizing was out of the question due to the age of the older animals.

Dad sat high atop the tractor, Parker drove one pickup and Keith drove the other. I was in the back of Parker's truck, a small Chevy, with a garden rake. The rake was not used to strike the yak, just to make the truck and I look larger and more intimidating. Doug was in the back of Keith's truck, a large Ford, with a smaller rake. What happened next was more blur and bump than I've experienced on any carnival ride. The pasture was not smooth, but typical Sussex County peat bog and rock. We ramrodded back and forth across the paddock. We were successful in getting the three females out of the paddock after a half-hour. The gate was closed and the three female yak were herded with the three vehicles down the path up the first hill. After clearing the hill the ladies saw the lush green grass of their new home and trotted right in. So much for the ladies.

The male of any species is more stubborn. We all went back into the original paddock in trucks with Dad on his tractor throne. After an hour of unsuccessful attempts, yelling, hollering, cursing at each other, bruised by the bumpy ride and our teeth and kidneys shaken by the rough terrain, it was decided that the old bull yak looked slightly winded. We all were exhausted but the yak was old, after all. It was generally decided to take a break.

Dad opted to stay on his tractor and "commune" with the old male yak. I told Doug to keep an eye on Dad. Dad had turned 68 years old that summer and I was concerned. Dad had also had many animal adventures in his lifetime that we laugh about now, but at the time were dangerous. Little did I know...

Parker, Keith and I went on about our daily chores that had not been completed for the day. Parker fed the carnivores (meat eaters), Keith fed the herbivores (grain eaters), and I was in the main building. I knew there was trouble when Keith walked in and said " You gotta see this!"

I looked out the back glass door of the main building and beheld a sight that chilled my soul. In the old yak paddock was my dad on his tractor. Sitting propped up on the hood of the tractor, his boot heels hooked into the hay fork on the front of the tractor and holding a ten foot, one inch diameter metal rod with a hemp rope lasso on the end, was my husband. The love of my life, father of my only child, was riding over the bumpy, boggy, rocky, log strewn pasture chasing a ton of angry yak, with my own father at the wheel. The pasture was wet and slippery to boot. If Doug fell off, Dad's 68-year-old reflexes combined with slippery conditions... or if Doug actually lassoed the beast it would yank him off the tractor, with slippery conditions...

I was livid. I knew immediately what had happened. Dad had suggested the action, and Doug, in awe of and relying on Dad's animal experience, had agreed, climbing onto a potential death trap. I didn't know who to be angrier at, Dad for suggesting or Doug for believing the rodeo antic would work. In Doug's defense, his expertise is in mechanics not animals, although he's learned a lot about animal behavior since. I was truly beyond livid. I let loose a slew of curse words that would embarrass the proverbial fisherman's wife. Keith didn't dare utter a word as he drove me rocket speed to the scene. Upon my arrival the two men gave up their Don Quixote quest, sheepishly kicking the dirt, shrugging, and stammering excuses like two little boys caught with their hands in the cookie jar.

The next day Parker suggested we box the old yak in his barn, lasso him in close quarters, and pull him out with the tractor. That worked, sort of. We had the old guy lassoed and pulled him with a larger diameter rope halfway up the hill on the path to the new paddock. The old yak still was not cooperating. We had trucks, fitted with cattle gates on the front, blocking the return path. Dad, on the tractor, was pulling the reluctant yak, the trucks bringing up the rear. We had to poke the stubborn yak with the handles of our rakes to get him to move a few more feet. Then the rope snapped. That rope was twice the diameter of the rope Doug had tried to lasso the old yak with the day before. I shuddered to think. The men put the trucks in low 4-wheel drive and pushed the reluctant yak to the top of the hill. Then, miraculously, the male yak spied his family in the vibrantly green, grassy paddock ahead and trotted to them of his own accord. Needless to say, it was a day I will not forget.

Doug and I had a long conversation about animal behavior after that move. I realized he did not know the animal behavior he needed to know to be safe around the zoo animals. It takes years to be able to read an animal's body language. You learn to watch the ears, eyes, and stance of an animal. The animal knowledge that had been given to me by my family's years of experience did not transfer automatically by osmosis to my husband. Sometimes it's hard to remember that.

The cougars were an easy move. Their entire den, with the cougars inside, was moved by tractor to the middle section of the zoo. In the spring of 1996 the men would reconstruct their natural turf enclosure in the new location.

After all the animals were moved and bedded down for the winter, the hard work began. We had to have all the animals, their enclosures, all wire fencing and posts off of Eric's property by December 1st, or face financial penalties. The men worked from dawn to dusk in cold weather, rain, sleet and snow included. I helped where I could. It was a very busy time for Doug, Jackie and I. Doug and I worked all day. Jackie was in a new school and quite upset about it. At night, Doug, Jackie and I would go to the house we were to move into, adjacent to the farm, to paint, clean and for Jackie to do homework. Jackie would often fall asleep on the floor while we painted. Afterward we would go back to Dad's house to sleep. The dawns came very early that winter.

By December 1st one third of the zoo's population had been moved and their enclosures dismantled. This was quite an accomplishment for a single month's work. On December 5th, Doug, Jackie and I moved into our house. The winter snows came as we settled into our routines. Jackie had some trouble adjusting emotionally to a new school and new friends, but was doing well academically.

In the winter we block off the auxiliary dining room in the main building to conserve on the heat bill. I set up a painting studio in the main restaurant near the windows with northern light. It was a warm and comfortable area. I am a people person, so the days without visitors were long and lonely, however the guys worked outside in the freezing weather, so I have no complaints. I spent the days painting signs and the miniature golf course inside the main building. I answered phones, scheduled my speeches for the next summer, and helped with whatever animal problems came up.

Visitors often wonder what we do all winter long while the zoo is closed to the public. Many zoos ship their animals south to warmer weather. Space Farms does not. Transport is always stressful to the animal. You never know what germs and diseases an animal may be exposed to in a foreign environment. We prefer to keep our stock on the farm to keep a close eye on them. Any animal that requires warmer temperature is moved across the street to one of the many barns on the property. All of the animals eat daily in the winter, just like in the summer. Each animal receives fresh water daily, just like in the summer. In the winter, bedding hay is replaced weekly or more often when the weather turns snowy and the animal tracks snow into its den.

The big cats all have dens attached to their enclosures. In the summer the dens are open on one side for the public to look in. In the fall these dens are closed up with only a small opening for the animal to come and go as it pleases. Stuffed with bedding hay, the cats stay warm and dry for the winter. The Atlas African lions, leopards, and jaguars acclimate to the colder weather, growing heavier coats as they would in the wild. Jaguars are currently found only in the warm rain forests of Central America. However, a few generations ago Lewis and Clark wrote of seeing a large spotted cat during their expedition in the upper Mississippi valley. Man has pushed this marvelous cat into the American southwest and the rain forest. It is capable of surviving in the New Jersey climate, as long as it has a den to get out of the weather and stay dry. Acclimation is a slow process. Zookeepers cannot bring a leopard, lion or jaguar from Florida and expect it to survive in New Jersey in the middle of the winter. If keepers bring it north in the spring and the animal has time to adapt to the slow decline in temperature through the fall, it will do fine with the proper accommodations, growing a warmer winter coat that is shed in the spring.

Fall and winter are a major project time at the zoo. I'm sure you can understand why we can't have heavy equipment in the zoo proper for small tykes to climb on while we are open to the public. Everybody knows how much little kids like tractors, bulldozers, backhoes and the like. That is an insurance nightmare waiting to happen. In the fall of 1995, we had moved one third of the zoo. Spring came and it was time to put together the enclosures we had moved in the fall. The woodchucks and prairie dog enclosures were reconstructed. Wire fencing was laid down, dirt for digging and burrowing piled high on top of the wire and the cement sides were raised into place. The painting of the enclosures was put on my long (and growing) list of things to do as the weather warmed up.

Chapter 30
Wild Turkeys

Dad has always loved wild turkeys. He was instrumental in helping to re-establish the wild turkey to New Jersey.

Wild turkey hens often make their nests in or by a hay field. The wild turkey hen is a dull, mottled brown color. Her camouflage protects her and the nest from predators. Accidents happen during the setting season, when the hen is incubating her eggs and does not leave the nest. This is in the spring (May) when farmers are mowing the first cutting of hay. Often local farmers would bring in a nest of eggs for us to finish hatching, after the hen had been accidentally mowed over.

We put the eggs in incubators and hatch the chicks. We are careful to raise the chicks with as little human intervention as possible to prevent imprinting on humans. Imprinting is the phenomena of an animal attaching to its adopted mother. An adult tom (male turkey) is aggressive during the breeding season. If that tom has imprinted on humans it will fight humans, having lost his fear of humans.

In all the years of raising orphaned wild turkeys, we only had trouble with two toms. They were living at the zoo when Doug, Jackie and I returned in 1995. In the summer months we shooed them outside the zoo's guard fences. In the winter, when the zoo is closed to the public, they roamed the zoo grounds. The wild turkeys would fly in on their own to eat the farm grown fresh ground corn in the hoof stock paddocks. The turkeys never bothered the men. Dad was particularly proud of these large 25-pound birds; they were excellent specimens.

The tom turkeys hated me (or perhaps loved me) because I was female. When I went out into the zoo they would follow me and try to jump on me. This may sound funny, but a 25-pound bird flying at you feet first can be very scary. Let alone dangerous. A tom turkey fights his opponents with his spurs, similar to a cockfight. During the breeding season the males fight to establish dominance. The toughest tom gets to breed the females. During the breeding process the male "drums" the female, "loving" her with his wings. This was the type of activity I had to deal with whenever the turkeys spotted me. Dad informed me that I had to be tougher than the turkeys. When they came at me I had to prove that I was the stronger turkey, to scare them away. Yeah, right!

Gramma had retired from the restaurant when she was 89 years old. Her little house was right next door to the main building. We all started every day at her house with a breakfast meeting to plan the day's activities and projects. Later in the morning we would stop by for coffee and cookies. Gramma had lots of friends that would stop by. Gramma was the stereotypical grandmother, with beautiful white wavy hair, roundish body and always in the kitchen wearing an apron. She was loved and respected by the entire community. She taught me all I know about raising infant animals.

One day my friend Judy stopped by and we were gabbing in the main building. It was about coffee time so Judy, who also adored my Gramma, decided to come to Gramma's for coffee. I took my hot cup of coffee with me as we walked the fifty feet to Gramma's little warm house. Halfway there, I spotted the turkeys running toward us. Their naked colorful heads were held out straight, I knew we were in for trouble. I am a hefty woman, but Judy is petite. I shouted to Judy, "Run to Gramma's, I'll hold them off!" Judy looked around and saw the advancing toms. She ran for Gramma's house and I was almost there when the first tom flew at me and hit me hard in the legs, jolting my whole body. Judy had made it to safety inside Gram's back porch and was looking out the storm door.

I poured my hot coffee on the turkey's head. He came at me again. A 25-pound turkey jumping on you, trying to drum you is scary and dangerous. I hit him in the head with my coffee cup. I could not let him get me down on the ground. He came at me again. I threw my coffee cup at him as he jumped on me for the third time. I kung fu chopped and kicked the turkey, trying to be tough, I'm sure I looked ridiculous since I had no formal training, only what I'd seen on TV. I was backed up against Gramma's storm door. I could not open the door for fear of the hormonally insane turkey following me inside and ransacking the house, not to mention hurting my 92-year-old Gramma or Judy.

I was getting upset and started to cry, the tom still aggressively attacking me. I screamed to Judy, "Hand me out Gram's broom!" Judy inched open the door and handed me my weapon. Gramma had heard the commotion and was watching out the door, cheering for me. I lightly smacked the turkey with the bristly end of the broom. Now both tom turkeys were jumping and strutting by me. The head honcho tom was jumping chest high onto me. All I could see was feathers and claws. I smacked him harder with the bristly end of the broom. That had no effect, no effect on the old turkey at all. I was scared, he had ripped my pants with his spur claws and was still coming at me, running and jumping higher every time. His buddy was helping out now, too.

Dad's voice crept into my memory. He had said, "You have to give them all you got." I got my muster up and turned the broom around. I smacked the turkeys on their bodies. The darned turkeys heightened their attack and came at me for another round. I hit them harder and harder. The toms still would not leave me alone long enough to get into Gramma's safe haven house. At this point Gramma and Judy were screaming, "Hit 'em again!" I got angry, I was not going to have my face or body scarred in a cockfight with a tom turkey. I turned the broom in my hands and used the handle baseball bat style. I slugged the head honcho tom right in the noggin. This was clearly self-defense. The head honcho tom staggered a few feet away and dropped to the ground. The other tom took off for the back of the zoo. I had survived, I was sweaty and shaking, but I was the toughest turkey.

I was perspiring quite heavy during my battle. I was also quite shook up, realizing I had hurt Dad's prize tom turkey. I staggered into Gramma's. Judy and Gramma had me sit down. Gramma whipped me up her recipe for all stress, tea with honey. I was cooling down when my brother Eric came in to visit Gramma and have some cookies. Eric took one look at me and asked, "What the **** happened to you?" I explained, tears still streaming down my face in battle shock. He then asked me if I had killed it.

"I don't know, the tom was laying in the backyard, last I saw him," I said. Eric went out the storm door and inspected the turkey, and decided the best thing to do was to put it out of its misery. I was still shaking when I drove to my house, a quarter mile up the hill. I changed my clothes and went back to the zoo to do the hardest thing I've ever done. I had to tell Dad I killed his pet prize turkey...

Dad was in the office filling out a dead deer report. I decided to make light of the situation, *Hey, it beats pleading for mercy.* I stuck out my hand and said, "Congratulate me." Dad looked over the rim of his glasses suspiciously.

He shook my hand and warily said, "Congratulations. What for?"

" I just killed my first turkey," I sheepishly replied.

"Did you hit one with your truck?" he queried.

"No, Dad, with a broom." I waited for the explosion, but Dad was not catching on.

"A BROOM? What Turkey?" he asked, as the possibility was dawning on him.

"Your turkey," I winced. Before Dad could catch his breath to chew me out, I gave a fast rendition of the story. To my utmost surprise, Dad bowed his head and quietly shook it back and forth.

"Come on, we might as well pluck it and I'll have him for dinner," he said.

I wasn't ready to be sorry about the turkey quite yet. I said, "Wait a minute; I want you to take my picture. You took a picture of Eric and his first turkey, and a picture of Parker and his first turkey, and their guns."

Dad looked at me incredulously. "You want a picture of you, the turkey and THE BROOM?" he questioned. I did not smile when I picked up and handed him my camera from my desk. We went out to Gramma's cherry tree where all the hunted wild turkeys that had ever been shot by the men of the family were traditionally hung. We hung Dad's bird on the traditional tree and took a picture of Dad's favorite, now dead, gobbler, Gramma's broom and me. Dad never said a word.

Later that afternoon I was finishing up a painting project in the main building while Dad plucked the deceased turkey in the restaurant's kitchen sink. I could hear Dad mumbling under his breath, peppered with a few choice curse words. I decided to ignore him. After he completed the plucking, he called me over. "What did you do to this bird, he's all beat up? He has a broken wing, a broken leg, is bruised all through the chest and his skull is caved in." I could tell Dad was upset about his pet.

The only defense I had was Dad's own words: "I poured coffee on him and he still came at me. I hit him with the coffee cup and he came at me again. I hit him with the nice side of the broom, and he still came at me. I hit him in the body, with the business side of the broom, and he still came at me. So I decided to be the tougher turkey and I hit him in the head!" I explained once again.

Dad solemnly shook his head and replied, "You're a turkey all right!" The bird was not wasted; Dad and Mira ate him for supper. I prefer store bought birds.

The state of New Jersey has laws against unlicensed individuals raising wild turkey chicks. This type of wild turkey attack is the reason why. Even professionals have occasional problem birds. Me? I still don't like turkeys—of any species.

Chapter 31
Three Little African Atlas Lions

In March of 1996 one of the female Atlas African lions that had been born at the zoo three years before had three cubs. This was her first litter. Often in the wild the new mom will lose her first couple of litters before she catches on to the idea of motherhood. In the wild eighty percent of lion cubs do not make it to their second birthday. We observed the mom to see what she would do. She was not actively taking care of the cubs. The grandmother lion was, but she had no milk. The decision was made to take the cubs and rear them by hand. Upon entering the back door of the birthing den, all the lions ran out. We knew then that we had made the right decision. Proper mother lions will not leave their cubs; they will fight any intruder that comes near the nesting area.

I was thrilled. These were Moses' grandchildren, my great grandchildren.

Dad helped me set the cubs up on the back porch next to the kitchen. We stuffed a large box with hay and put a covered playpen with a latch on it to keep the public from touching them. We posted a sign saying what they were and not to touch. The young cubs drank greedily and did well for the first few days. I took the infant cubs to Gramma's house every day for the breakfast meeting of the family. Gramma loved all the babies at the zoo. Gramma's body had grown old and frail so I had to make sure the cubs did not scratch her and cause her elderly immune system to react. At 92 years old, Gramma could not physically care for animal infants any longer, but the twinkle in her eye, the constant concern and advice let us all know exactly who was the expert in animal infant care.

I had a speech to give on Saturday and decided to take one cub with me. Dad agreed to feed the remaining two cubs while I was gone. The babies were strong and healthy, eating well and pooping appropriately. I was back from my speech within three hours. I noticed a pad lock had been placed on the door of the playpen. I sought out Dad and asked him for the key. He explained that some jerk (a zoo visitor) had opened the clasp on the door, picked up the two cubs and was passing them around to his family and friends who were visiting the zoo with him.

I was livid. Dad and I both knew the reason for the Do Not Touch sign. It was not that the toothless two to three-pound lion cubs might harm someone, and we might get sued. The warning was because people carry germs from their pets on their clothes. House cats and dogs carry certain germs that do not affect the domesticated animal. The domesticated animals have built up a tolerance over generations for those germs. Those domesticated germs are lethal to exotic animals.

I moved the cubs into Dad's office next to the window and locked the office doors. Dad tried to convince me not to worry, but I noticed him checking on the cubs more frequently than normal. Within twenty-four hours the first male cub was failing fast. I called Dr. Ted Spinks, our veterinarian. He was on his way. Like most zoos we keep a vet on retainer. Dr. Spinks is great and he is very attentive when we call. He knew we had a problem when I explained what had happened and the symptoms of the cub. He instructed me to keep them separated till he got there.

The little male cub died in my arms moments later. I was heartbroken. I went to Gramma's house and explained to Gramma what had happened. She was sitting in her favorite rocking chair and I put my head in her lap. She petted my head with her arthritic gnarly hands as I sobbed my eyes out. Gramma reassured me that I had done everything the best I could. She reminded me that without us all three cubs had no chance at all. After a half-hour Gram had settled me down, her experience and understanding soothing my conscience. Gramma washed away my fears with her traditional salve of tea with honey. She had also loved and lost animal infants and could empathize with my affection for the tiny lost lion. Her absolution cleared my soul. Gram told me to splash some cold water on my reddened eyes and check on the other two. Dr. Spinks came and we started intravenous fluids on the female cub that had started to go down hill. He left me with a supply of saline and antibiotics to help the kittens through the night. The second cub died at midnight.

The third cub appeared to be doing well and had not succumbed to the listlessness that the other two cubs exhibited before their demise. I was hopeful that the cub that I had taken with me to my speech had not been exposed to the germ that killed the other two. Gramma and the Vet were encouraging.

After twenty-four more grueling hours, as suddenly as the others, the last cub was in decline. I eyedropped fluids, and fussed over her to no avail. I was a failure. The cubs I had waited twenty years to raise were dead.

I cannot describe the depths of despair I experience when I lose an animal infant. I feel so unworthy of the trust that God and my family have put on me. I sought out Gramma and her tissue box again. Her ninety-two-year-old eyes cried with me and for my pain. I remember her words to me when I told her I had failed. She said, "God will use you again and again, because He knows you did the very best you could for His babies." Often her voice still comes to me and speaks the same wisdom. Sometimes it works. I miss her shoulder to cry on.

Chapter 32
Guns and Brass

Every year I would give speeches on Space Farms Zoo and Museum. The speeches were free, promotional and educational programs tailored somewhat to the audience I addressed. That spring I was scheduled to speak to the retired police officers of Sussex County. At that particular time I did not have many babies that were transportable, only an owlet and my favorite snake, Charlie the king snake. Since I was taking animals that were not messy, I wore a nice dress with a blazer over top instead of my Space Farms uniform.

I arrived a few minutes early, as is my custom, to get the layout of the room. There were about fifty retired officers, jolly and relaxed, maybe a little bawdy, but very friendly. I recognized a number of the men as friends of the family. I was the only woman in the room aside from the waitresses. I ate dinner and conversed with the officers around me. Then it was time for me to speak.

I have a policy of not drinking alcohol (I don't drink much anyway), before a speech. I like to be on my toes and work the audience. My policy did not apply to everyone else in the room. The men were attentive and asked a number of questions about the animals. After the animal section of my talk, I pulled out the antique guns I had brought from the Space Farms Museum. Now I had the full attention of every officer in the room. I had brought an antique powder horn, engraved with the map the soldier had carved in it. It was from the French and Indian War, a nifty and unique piece. I had brought a number of muzzleloaders, explaining how they were loaded with patch, powder and balls. I went down the list of interesting facts about each gun, just as Dad had prepped me that afternoon. I exhibited a number of antique bullets.

I had hunted as a youngster, but had not picked up a gun in 30 years. So I was a little nervous speaking on a subject that everyone else in the room was an expert on—guns. I went over rifling in the barrels, powder, the size of the powder hole, consistency of quality of powder available in the old days, patch boxes in the stocks, type of wood in the stock, etc. I had taken copious notes as my Dad had explained each gun to me. I parroted the information to the retired officers. I only had a minimal understanding of everything I spoke about. In short, I was faking it. The men in the room soaked up every word, so I must have been on "Target." My Dad is a great teacher.

After I had depleted my notes, I asked for questions. I prayed I would know the answer. A few minor questions I fielded with ease. The last question was a knockout. A younger officer asked, "What type of permit do you need to carry antique guns?" Bang! I knew I was dead in the water. I had no clue. It briefly flashed through my mind that it was *9 pm. Dad was asleep. Doug was home with Jackie; Parker was home with two little kids. Even the streets of Beemerville were rolled up at 7 pm! Who could I call to bail me out of jail this late at night? I was crazy to bring two long rifles and two pistols in my tote bag (concealed weapons) without some sort of permit. I never even thought to check, I did not make one phone call. I was going to jail for sure.* The room was deathly silent. The waitresses stopped their bustling and held their breath with me...

"Well," I stammered aloud, as I thought, *Girl give this all you got*. I stood up straight, hands on hips, parting my blazer, showing off my 38Cs, "I have no balls..." (*Pause, pause, let them have their laugh*), "no patch and no powder, which are the ingredients needed to fire these guns. These are unloaded guns and I do not carry the appropriate bullets." The room roared with the raunchy laughter I had hoped for. "I do, however, pack a full set of 38s and," (stand up straight, *pause, pause*), "45s," I said, slapping myself on the hips. The entire room and the waitresses broke up with laughter. The moment passed and I did not go to jail for carrying concealed weapons. *Whew, give me a room full of retired, relaxed police officers any day!* I have bumped into many of the same retired officers since, and am known affectionately as "The girl with no (muzzleloader) balls."

Chapter 33
The Leopard's Eye

Space Farms Zoo, under my Dad's direction has established longevity records for the bobcat, jaguar and puma (cougar). Many of our residents live long, long lives. As animals get older, they face many of the same old age problems that we humans encounter as we age: Arthritis, cataracts, achy bones, loss of hearing, loss of hair, loss of continence and loss of sexual drive are all effects of aging that occur to animals also.

In the spring of 1996 our black leopard female was eighteen years old. That is getting up there for a cat. Most cats live 18 to 22 years. She had developed cataracts in both eyes, one worse than the other. Dr. Spinks and the family had discussed the pros and cons of an operation for the cat. The older the cat (or human for that matter), the harder it is for the animal to come out of anesthesia. We had decided to let her live out her life without taking the chance of putting her into a sleep she may or may not come out of. The black leopard had memorized her enclosure, could find her food by smell, and seemed to be fine in every other way. She had some vision, but we did not know how limited. She showed no signs of arthritis, or kidney problems. She spent her old age sitting in the tall grass, basking in the sun wiggling her tail, and sniffing the breeze. She seemed happy.

When the cougars were moved to the new section of the zoo, a new den had to be constructed for them. After designing and planning of one new den, Doug and Dad decided to replace the other big cat dens. The dens were fifteen years old and needed more repair than they were worth. "If you are doing one, you might as well do the other two," Doug explained. So three new cat dens for the cougars, leopards and jaguars were built. Each den was constructed of a concrete base (for cleaning purposes) and a shelf so the cats could rest on a wooden platform that was stuffed with hay in the winter. Three wire cages were built on the inside of each den so each of the two cats in each enclosure would have their own "space." We could walk in the den buildings themselves if we needed to check on a big cat for whatever reason. In the summer the front of the dens were open. In the winter, the fronts of the dens were closed to keep in the warmth. Doug designed the new dens. Doug and Dad went over and over the plans, perfecting each section.

Finally the dens were built, they looked great, and the cats moved in picking their own rooms inside the dens like children exploring rooms in a new house. I painted them barn red in keeping with the farm theme.

Three days after the buildings were complete and the cats had "moved in," our old female black leopard showed up with an eye injury. Part of the lens was protruding from the eyeball itself. We did not realize how blind she had become. Evidently she had memorized the construction of her old den and with the construction of the new den, she was lost. The black lady leopard had bumped into something and injured her eye. We called our Vet. Dr. Spinks and Dr. Servideo came right away. After examining the old female from a distance we all discussed the problems with anesthesia on geriatric cats. Left alone, the eye would become infected. Or we could just put her down. She was a sweet, old cat, still up and around with no other old age problems that plague cats. Dr. Spinks said there was another option. He could operate and take the eyeball out, but she may not come out of the anesthesia. Dad, Parker, Doug and I discussed it and decided a chance was better than no chance at all.

Dr. Spinks administered the anesthesia with a punch pole and we waited and watched the distinguished old lady's breathing. When she fell asleep, Parker and Doug went into the den and brought the leopard out on a stretcher. The stretcher was put on the back of one of the zoo's jitneys. Dr. Spinks went to work. I was fascinated to watch him deftly scoop out the damaged eyeball with what looked like a melon scooper. Next he cut the rim of the eyelids off to create an open edge. Dr. Spinks sewed the two edges of the eyelid together. "When the wound heals," he explained to us, "the eyelid will heal shut, so she doesn't get a dry socket or other infections in the orbit." Made sense to me. "She'll look like she is winking." Dr. Spinks continued. "I've done this on housecats and it turns out well." Dr. Servideo administered antibiotics and gave instructions on how to inject the leopard's food with oral antibiotics for the next ten days. The geriatric cat was placed back in the den in the room she had chosen. The young male companion leopard was locked out, he could come in to see her but not come in her room. We watched and waited for the old girl to wake up. We all crossed our fingers and monitored her breathing. An hour or so later, she was holding her head up, winking at us. She lived two more years after that, dying peacefully in her sleep during the night. I never thought she'd come out of the anesthesia. I'm glad I was wrong.

Chapter 34
Norman the Holstein Calf

Space Farms Zoo and Museum is technically a farm. We grow all our own corn, hay, alfalfa and oats. The name of the farm originated from our family surname, Space, and the multiple farms my grandfather and father bought and added to the original 1/4-acre purchase in 1927. Hence the name, Space Farms Zoo and (later) Museum. The 100-acre zoo property is open to the public, the other 350 acres is farmland, wetlands and open space woodlands. Because we have the word farm in the name, and the fact that we are descendants of farming families, we kept farm stock at the zoo. We like the chickens, pigs, geese, and ducks, as does the visiting public. The assorted farm stock just seemed to gel with the country atmosphere at the zoo. Every year we borrowed a calf from a local farmer. Everyone expects to see a calf at a farm.

The first Holstein calf I raised at the zoo as a zoo animal (I had helped in the Space Dairy Farm as a youngster) was Norman. Norman was a black and white Holstein dairy calf named after the calf in the movie *City Slickers*. Calves bond readily to the person who feeds them. Norman was no different. I would walk out with his bottle every morning, call his name, "Noooorman, Noooorman." He would bellow and come running, drooling along the way in anticipation of his bottle. Farmers would feed a bull calf for a couple of weeks and then ship it off to market. You and I call it veal; the farmers call it business. Anyway, the bull Holstein calf grew and grew on his special formula. The same formula that farmers fed their calves for only two weeks, I fed Norman for the entire summer. He put on a lot of weight and was a substantial sized bull by fall. But, still my big baby. Norman was four foot high at the shoulders, head above that. He ran in the grassy treed run alongside the Waterfowl Lake at the zoo. Norman was in good shape. Every day he would run and frolic alongside the lake, chasing geese, ducks and run to the visitors who brought him corn to eat. Twice daily he would get his 1/2 gallon bottle full of special milk.

Fall came and it was time to send Norman back to the farmer we borrowed him from, who assured me that Norman would grow up and be put to stud. I had quizzed the farmer and I'm sure he sensed my apprehension, as he assured me again that Norman would be put to stud. He knew I did not want to hear any other answer. The farmer showed up one day with a pickup truck, a stock trailer, two extra burly men and they were ready to load Norman up in the trailer. When the farmer pulled out a rope and said, "Ready, guys?" I realized they intended to wrestle the huge calf into the trailer. I shouted, "Wait a Minute!" and ran with my bottle. I explained to the farmer that I would gently walk Norman up into the trailer as he drank the bottle. Three burly Sussex County farmers looked at me with a chuckle in their eyes. "Yeah right, Lori," they said, and I said, "Let me try, if it doesn't work then you can wrestle the 300-pound calf into the trailer."

Norman was a distance away, perhaps sensing the coming change in his home field. I called him, "Noooorman, Noooorman." As per his Pavlovian training, he came running. I started him on the bottle and began to walk toward the trailer. Norman sucked on the bottle. I walked out the gate and up into the trailer with Norman still on the bottle, ambivalent to all but the bottle. The farmer smiled and started to put a halter on Norman's face. We threaded the halter over the bottle and secured Norman to the inside of the trailer with a rope on his new halter. Norman still drank, oblivious to the change in his surroundings. I let him finish the bottle.

After closing the door to the trailer, the three husky farmers were still chuckling to themselves, shaking their heads. It was such an easy load up they were amazed. Ever since I've raised calves for the same farmer, and when it's time for the calves to go, they stop and get the calf's bottle and me. We started a tradition; the calves always get to finish their bottle in the trailer. The local farmers don't laugh at me anymore, but I don't let them see my teary eyes as they pull out the gate with my baby bulls either.

Chapter 35
Three Little Stinkers

A cardboard box on the office doorstep first thing in the morning is always interesting. Maybe it is a shipment of parts for the farm vehicles, some foodstuff for the restaurant, souvenirs for the gift shop, or maybe, just maybe something for my department. Everyone loves a surprise package.

That spring the box was for me, Dad and I could tell by the folded top and the sound of scurrying feet inside. Carefully we opened the plain brown box. We did not need to peek in after the odor wafted past our noses. Dad chuckled and said, "These are all yours, Lori!" Three little skunks peered up at me, beady black eyes shining. Dad and I talked about the possibility of rabies. Dad declared that the babies looked fine, but that I should be careful not to touch saliva or get bitten for the next ten days.

They didn't smell all that bad. Skunks don't spray till they are older, or so Dad reassures me. They were obviously babies, we estimated they were three or four weeks old. Taking no chances that the three little stink bombs would go off due to fright, I ever so carefully moved the box to the barn across the street. I set the infant skunks up in a small cage. Every day I fed them with a small kitten bottle and Esbilac. Moe, Curly and Joe drank a few drops at a time every couple of hours. Gently, patiently, and cooing soft words, I handled the skunk kittens, until they got to know me and were no longer frightened of me.

After ten days of worrying about the possibility of rabies, my totally rational germaphobia passed. By this time I was able to handle the kittens easily, not overly worrying about the skunks spraying me. Moe, Curly and Joe had adopted me as their mom. I would take the three stooges out for a walk on the lawn and let them romp in the same flower garden the chimps used to eat flowers from. We had a great time. All the employees knew the skunks were not deodorized yet and would leave us alone. The four of us would walk around, tails in the air (only theirs), enjoying the green grass. The babies would dig for grubs and insects, part of their natural diet. They grew gracefully and were adorable fluff balls by six weeks old.

Dad and I had discussed deodorizing the three and keeping them in the zoo. The skunks had bonded to a human and would not be afraid of humans or cars, because of their exposure to us. All three were males so we did not have to worry about birth control. Why worry about skunk birth control at a zoo? Even though Dad could deodorize the skunks, any progeny they produced would be fully functional skunks. All it would take is a visitor to scare the potential new babies, and kaboom, the skunk bomb would go off. Can you imagine the smell? Can you imagine the lawsuit! So it was a good thing they were all males.

On one of our daily walks, a passing truck blew its horn to say hi to me and frightened the young skunks. Moe, Curly and Joe scooted under one of the employee's cars parked nearby. Tails held high and hind feet in steady spray shooting position, they were ready to take on any threat. I tried to call them out. No luck. I tried to entice them out with food. No luck. Every time I would run to one side of the car, they would quickly play round robin and take their stand on the other side. There was only one way and only one person who could do the retrieval and that was I, the Skunk Mommy. I am not a tiny woman. I wedged myself under the car and once again in my life came face to face with skunk butt. I called them again. Responding to the voice of their mom, and feeling safe in their newfound haven under the car, the three troublesome kittens came to my side.

Oh great, I thought, *now I am wedged under the car with three live stink bombs on two sides of me.* It must have looked quite silly. I wiggled out from underneath the car with a captured skunk in each hand. Talk about loaded weapons! I dumped them gently but speedily, into the transport cage. Joe came running to be with his brothers. I put Joe in with Curly and Moe. After putting the soda can sized skunks back into the barn, I decided I would not take them out for a walk again until they were deodorized.

Most skunks give birth in May in New Jersey. Years before, when we were raising skunks for Dr. Hagadorn, Dad had always deodorized the young around the fourth of July. We scheduled the operation for the days following the fourth, just in case something went wrong and the skunks sprayed; we wouldn't stink up the farm for the holiday. Skunk juice is very potent; a teaspoon can smell up a square mile. Dad had me assist once again in the skunk deodorizing. We performed the operation behind the barn, in the open air, just like years before. I was impressed with my dad's gentle, yet precise, expertise while operating on the skunks years ago and appreciated it more now that Dad was so much older. His hands were still steady, performing the delicate operation. We wore our worst clothes; needless to say they would be thrown out later. We looked a sight in our raggedy clothes, our worst for the wear hats, latex gloves (a new invention since our earlier operations) and space man goggles. The operations went smoothly and the kittens were placed on a soft diet. After twenty-four hours they were their normal perky selves.

    A six-week-old skunk kitten is adorable. The eastern skunk is black with a white stripe crowning its forehead, down its back and coming to full bloom on its tail. Intelligent as a cat, inquisitive and affectionate, the skunk kittens were lots of fun. You can house train a skunk to a litter pan. Parker had one as a pet in the office for years. Parker's pet skunk, Channel, lived under the safe. Until a few years ago it was legal to have a skunk for a pet in New Jersey. I did not house break these three since I knew they would live in the zoo, on natural turf.

    A trip to the Vet for shots and Moe, Curly and Joe were able to go on the lecture circuit with me, after a couple of shampoos. I took the kittens with me to give speeches. If you ever want to clear a room, pull a skunk out of a box. Lecture after lecture, it happened every time. By the third time I took the three sweeties out with me, I had learned to warn the audience what I was pulling out of the transport box and the fact that they were altered so they could not spray. Some say you can still smell the skunk smell after the deodorizing, but I never could. I have a very sensitive nose, or maybe my heart is bigger. The three little males were put in the adult zoo that fall, and have been sleeping the days away ever since. Skunks are nocturnal after all. Not everything can be adjusted with an operation.

Jackie loved to play with the little striped skunks.

Chapter 36
Dead Snake on the Road

I traveled up to an hour and a half to give lectures to promote the zoo. One campground I visited was quite a ways away. I had packed my museum artifacts and animals (one mammal, one bird, one snake) and hit the road. Campgrounds are usually located in the back woods, accessed by long windy, bumpy roads. I gave my speech on the skunks, and the owlet I had brought with me. Then it was time for the snake. I had chosen one of the milk snakes from the den, again.

I precede my snake lecture by saying, "I'm now going to bring out a snake. I am not here to scare anyone. If you would like to touch the snake please hold up one finger on one hand." As I am saying the words, I am untying the knot in the pillowcase/ snake bag. After the oohs and aahs, I give my snake speech walking around letting people touch the snake. I would hold the head of the snake away from the people and let them touch the lower body and tail section. This would prevent the public from getting bit if someone grabbed or frightened the snake. I had followed this procedure successfully many times. Not this time. As I untied the knot and looked into the bag, I was horrified to see my snake was dead! Dead as the proverbial doorknob! The milk snake was in a perfect coil, but upside down. *What to do? What to do?* He was my favorite milk snake.

Fifty people had their eyes trained on me, waiting to see the snake. This was an extremely bad situation for public relations, and Dad had given me his public relations position, too! Luckily, the snake died in a coiled position, so I flipped the bag a couple of times till the snake was upright. Then I scooped the snake up in the palm of my hand, holding his head up in the traditional jaw hinge hold. I gave a twenty-minute speech on snakes with a dead snake in my hand. Only one man in the audience asked me if my snake was sick, or perhaps going to shed soon. I said yes and packed the dead snake up. I avoided the snake petting section of my lecture and put the snake back in the bag and tied a knot in the bag when I was done. That was a difficult situation, I could have starred in a deodorant commercial.

Chapter 37
Yankee Doodle Dandy, the Lion

Every day is busy at the zoo. The guys start right off cleaning animal enclosures at 7 am. I'm busy with the morning feedings in the nursery. Every year we get a token calf from a local farmer and it is my job to feed it a bottle. We always have white tail fawns on the bottle, Mouflon sheep, farm goats and whatever else was in need of human intervention. It takes me about an hour and a half to get the nursery cleaned, fed and watered every morning. It takes the guys that long to clean their sections of the zoo also.

The fourth of July is always a busier than normal weekend. The day dawned bright and sunny, we knew we would have lots of visitors. The men and I had finished our morning chores. Keith, one of the zookeepers, was coming down the path through the zoo on a John Deere tractor. The speed of the John Deere drew my attention as I rounded the corner of the main building carrying buckets of manure from the nursery. (Yes I do that too!) Keith screeched the tractor to a halt at my toes. Keith's red curly hair glistened in the sunlight, wildly escaping from his baseball cap. The sparkle on his hair matched the sparkle in his eyes. He winked at me as he said, "Guess what I've got for you?" I was in no mood to play games with the young keeper. Didn't he know it was 4th of July? We were going to be busy and I was running late on my chores.

I was still trying to figure out how to say nicely what I wanted to say when he opened his shirt. Tucked warmly between his Space Farms work shirt and his T-shirt was a little Atlas African lion cub. I was euphoric. I took a step up onto the tractor and transferred Keith's enormous smile and the tiny lion cub into my possession. I glanced toward Gramma's house with the thought, *You're right Gram, and God is giving me another try. I will not fail.*

Keith had found the lion cub on the grass outside the protective guardrail. The cub had somehow gotten out of the den, past the fifteen feet concrete shaded section of the enclosure and crawled onto the sunny lawn. Dad and I discussed putting the male cub back in with the mother. It was decided that if the second sister, the new mother, was so inattentive as to let her newborn cub crawl twenty feet out of the den and let the cub fall off the cement side into the grass, she was not going to take care of him. I had another baby to love.

Yankee Doodle Dandy Lion was born on the fourth of July. Yankee was a good-sized cub, weighing in at two pounds two ounces and 14 inches, nose to tail. I immediately placed him in isolation in Dad's office window. A heating pad covered with a soft towel from my house made a soft, snuggly bed. A recently washed stuffed animal (a giant Snoopy) was placed in the box as a comfort/security toy. The newborn cub was the size of a woman's size 6 sneaker and looked so small next to the eighteen inch Snoopy.

Lion cubs, like some other newborn animals, will nurse till they get sick and throw up if you allow them to drink all they want. In nature, mom does not sit around allowing the infant to drink continuously. Mom lions are constantly moving, twitching, licking the infant, restraining the amount of time the cub has to nurse. In litters with multiple infants, the siblings compete for the nipple, the weaker cubs being pushed aside. Often these weaker infants will fade away and are removed from the litter by mom. Eighty percent of lion cubs born in the wild do not make it to their second birthday. Yankee drank voraciously the two tablespoons (one-ounce) of Zoologic 42/25 formula I allowed him to have. This formula is 42% protein and 25% fat, similar to cat's milk. Yankee's appetite was a good sign. I checked on him constantly, not getting much other zoo work (painting, cleaning, etc) done for the first few days. The guys chipped in and covered "my other work" while I was consumed with Yankee and the other infants in the nursery (nine white tail fawns, a calf, three skunks, two sheep, and two goats, all on a bottle five times a day, and one fuzzy owlet).

My family had always teased me about being a germaphobe. Well, if they thought I was bad before, they thought I was truly nuts now. I was obsessive about my new baby lion. I would not fail again. Yankee was kept away from the public. Visiting public could only view him through the glass windows. Dad would babysit if I needed him, no one else was allowed to touch (or breathe on) the little cub. My husband, Doug, and daughter, Jackie, did not touch him for three weeks. Jackie anxiously waited the day that I deemed the cub strong enough to be handled. Newborn cubs are very much like human infants. They eat, burp, pee, poop and sleep. Then repeat. That was the schedule I followed for the next two weeks. Every two hours I was feeding, stimulating for urine every meal and bowel movements at least once a day, then letting Yankee drift off to lion la-la land. In between over and around, I fed and cleaned the other animals in "my" nursery.

Yankee's eyes were still sealed shut when I received him. The first eye opened at five days old, the other by day seven. Cubs always look cute during this time, like a plush stuffed animal trying to be tough as a pirate. Our lion cubs have foggy ocean blue eyes when they first open. Slowly, by four weeks, the eyes turn sharp grass green. As adults the Atlas lions have honey golden eyes, intent, and clear of focus.

Yankee grew fast; by the end of two weeks he weighed four pounds, five ounces and was staggering on his feet trying to walk. Infant lion cubs are funny when they first start to walk. No concentration, or fear of failure, they swagger along, like a fat old cow. Four or five steps then Yankee would lose the concept of walking and nosedive into the carpet on my living room floor. No harm done and still intent on his objective, usually coming to a bottle, or me for loves, he would start all over again. First the front feet were put into position, then slowly; methodically the individual back feet would be swerved into place. Once all four feet were in just the right spot, Yankee would haul his round little belly up off the floor. His tummy was the most beautiful thing I had ever seen (next to my daughter of course). I worked very hard to put that belly on him. The healthy round ball of his belly weighed him down, giving him a swayback appearance.

All day long he was in a playpen in the office. At night Yankee waddled through our house, practicing his swagger. By three weeks old he was an accomplished walker and starting to trot a few fast steps. I worried about him constantly. There is only one thing worse than a protective mother lion, and that is an over protective adoptive mother lion. Jackie was given the ok to play gently with the little lion, after his first set of immunization shots. I also took Yankee to visit Gramma at her house, which was adjacent to the zoo proper. Gramma delighted in the visits. She had raised many zoo babies herself and knew the love I felt. The more people the lion cubs are exposed to, the better for the friendliness factor. No one wants a lion that only "works with" one person, especially if that person is not you.

As all other working women, my after work time was full of housework chores to do to keep my small family fed and clean. While I did the chores, Jackie would "babysit" the swaggering lion cub. Because Yankee was kept safely in the playpen while I worked in the daytime, his exercise time during the evenings was very important to his development. Jackie played with Yankee, rolling balls, crawling around on all fours, playing patty cake, and rubbing him vigorously all over. While I was busy helping Jackie with her bath and bedtime chores, Yankee was supposed to be resting. I often returned to the living room to find Doug on the floor letting the lion cub crawl all over him. Yankee was all mine after 10 p.m. That was my special time, the time I could just lay on the floor and cuddle the chubby playful cub. Those were the minutes that I treasured after a fourteen-hour day.

A mother lion gives her cubs a good work over with a large sandpaper tongue, often pinning her cub down with a massive paw. It is important to roll the cub onto its back as a show of superiority. The cub on its back is in a submissive position, instinctually. Performed with love and in "play" mode, this re-enforces "mother is in control" instinct. If and when you ever pick out a kitten or a dog as a pet, this simple test will tell you if the animal has been handled properly, i.e. with love, and the animal will trust you to be the master.

Another job of Jackie's, at this stage of Yankee's development, was to keep an eye on him in case he started to squat to pee or worse. At three weeks old any stimulation of the bottom side can result in... well... results. Climbing over a pillow, a mountain of blankets, or even over our bodies stimulated urine movements. This type of exercise was needed for normal muscular development and coordination. Up until this time I had tickled Yankee after each feeding on newspaper. At three weeks I started to take him outside on the tall grass with a leash on his baby sized collar. I was not using the leash to restrain Yankee, just to catch him if he decided to take off.

During this time we quickly learned to pick up any newspapers lying on the floor of the living room. We also had to lion cub proof the house. Crawl spaces behind the couch and TV were blocked off. Curtains were raised above the cub's reach. Our housecat, Daphne, watched the little lion cub with interest as her food and water was placed above the cub's playful reach.

At two weeks old Yankee had gone to the Vet for his first set of shots. I still worried about germs, but it was time at four weeks for him to go outside to the zoo nursery. Dad and Doug built him a 20-feet-long by 10-feet-wide play area with a top to provide shade. Yankee loved being outside. Wouldn't you know, his third day outside, Yankee came down with a fever and was not sucking as voraciously on his bottle. I was mortified. I called Dr. Spinks' office and made an emergency trip to the Vet's office. I imagined the worst. I felt quite the silly mom when the Dr. Servideo (Dr. Spinks' associate) said Yankee was simply cutting his carnassial teeth! Now I know every pediatrician will say that children don't pop a fever when they cut teeth. My Gramma and every other experienced mother will tell you different. I should have listened to my Gramma. The next day after two needle sharp teeth poked through Yankee's gums, the fever was gone, and he was his happy playful self. As in nature, it was time to start Yankee on meat. We always had a good supply of venison. My Dad would go out daily and bring back fresh road kill deer for the animals in the zoo. Yankee was given boiled bones with some meat still on to chew. This helped him to cut teeth through his gums. I would also hand grind raw venison to the consistency of hamburger for him to eat, starting with a walnut sized meatball and gradually increasing the size weekly. At five weeks old Yankee had all four carnassial teeth and the small needle sharp incisors between those big teeth that lions are known for. I was amazed one day to feel Yankee's new molars with my fingers. Lion molars are not flat on the top like all omnivores or herbivores. Lions are carnivores, and as such have mountainous razor blades for molars. Most carnivores rip, tear and shear off chunks of meat and swallow with very little mastication (chewing). They shear off a chunk of meat and toss their heads backward getting the meat to the back of their throat to swallow. If you think about it you've probably seen dogs eat that way. An adult lion's jaws are very powerful, able to snap the major bones in an antelope or deer without effort.

Part of the training procedures I used with Yankee (and all other cubs to come) was to put my fingers in his mouth. This familiarization process was important so the vet could easily check his mouth, teeth and gums. This process is started early. As a newborn, with the first insertion of a finger to suck on and then a bottle nipple I would gently say, "Open." The open command was repeated every time a bottle was given, or when I put my fingers or meat into his mouth. I would say, "Open" to inspect his mouth and teeth, putting my fingers inside his mouth. Every time he was rewarded with food or loving pats and kind words. Eventually he would open his mouth on the command. This simple procedure was important for his future. It is easier to get a lion to open his mouth with a loving command than to try to force his mouth open. When training a large animal you must always remember that someday he will be stronger than you will. Why must a lion open his mouth on command? As an adult you must check their teeth for problems (lions can get cavities too) or occasionally give medications. If you can get them to open their mouths willingly, you can use a syringe or squirt gun to get the medication in their mouths. This process is a lot easier than having to tranquilize the animal for a good health check up.

Yankee grew quickly and was friendly and happy. He continued to eat more and more meat, and was cut back on bottles. By ten weeks old he was on three bottles a day, and eating an eighth of a deer daily. Occasionally I would give him chicken or beef, but he preferred venison. His paws were massive; he was going to be a big cat. At ten weeks old he was already double the size of a full-grown beagle. Yankee was a lover, a terribly affectionate huge kitten. He would of coarse get frisky now and then; after all he was a lion cub. I was madly in love with him. Every night he would still come home with us. Jackie and Doug would spoil him as much as I. Often I would find Jackie and the cub curled up asleep together on the floor or the couch. Yankee was the only animal I allowed on the couch. He was a great pillow, and would warm my feet, only occasionally sucking on a toe if he wanted a bottle. At bedtime he would go into a dog show pen I had set up on my enclosed front porch. He had his own comforter, pillows, and stuffed plush toys.

One day Dad took me aside and said it was time to put an ad out in the zoo papers that we had a lion cub available. I was broken hearted. I realized it had to be, that we could not keep Yankee. We had three lions. The father lion would kill Yankee if we put him back in with the pride. There can only be one King of the pride. To build a separate acre pen for a cat we did not need was ridiculous, business wise. I worded the ad myself and told Dad that I wanted to interview the perspective new owners. Dad knew I was emotionally attached to the young cub and as any understanding father would do, agreed, with the understanding that Yankee was to "go" before winter. It was only September. I think Dad knew I would procrastinate.

In order to distance myself emotionally, I attempted to leave Yankee in his play yard in the zoo nursery at night. The first night, I couldn't stand it any longer than dusk and brought Yankee home to my house. Doug smiled and shook his head. The second night I did better. I left Yankee till my bedtime at 11p.m. It's hard to get to sleep while imagining your baby crying at the zoo. Yankee was not crying, he was asleep, so I woke him up and brought him home. I was there anyway, so I might as well sleep better, I rationalized. The third nighttime was the charm; I took a flashlight and checked on him. Yankee was sleeping at 11p.m. So I left him at the zoo. It was tough, but it had to be done. I was at the zoo early the next morning, in my pajamas at daybreak, just to check on him. Then I came home and got dressed for my workday at the zoo.

The word was out to other potential zoos, trainers and sanctuaries. It was required that the potential new parents have all the proper permits and papers according to the laws. The USDA, the State of New Jersey, and the state of the prospective new mom and dad all have laws concerning the possession of exotic species. Then there are Lori's Laws. I did not want Yankee to be placed in a cage and forgotten for life. I agreed to the third phone call and let the perspective mom come for a visit to see Yankee, and more importantly, for me to see her.

I was pleasantly surprised and adequately impressed with the instant affection between Yankee and his soon to be new mom. She was a private owner with a batch of tigers. She wanted to branch out into Atlas African Lions. She knew her facts, showed me pictures of enclosures and explained that she had an experienced private keeper just for her cats. I decided she would give Yankee the home and attention I wanted him to have. I let her take Yankee home to Pennsylvania on the condition that I would have visitation rights. We signed all the legal paperwork. Yankee left with her. I went back to work that afternoon with tears welling out of my heart and streaming down my reddened face. I avoided everyone. All the men at work knew and avoided me.

I kept in touch with Yankee's new mom for a while. She invited me to come visit. I never did. If I visited him and he cried for me, I would be upset. If I visited him and he had forgotten me and attached to his new mom, like I wanted him to, I would feel abandoned. It is a no win situation. I was glad to get progress reports and can say that he is living the life of a pampered pet, the last I heard at Christmas.

I would often find my babysitting daughter, Jackie, asleep with Yankee Doodle Dandy Lion at the end of the day.

Chapter 38
Salutations of Holiday Cheer: A Note to my Friends

Hello, salutations of Holiday cheer
I just can't believe it's been a whole year!
The spring came and went with babies galore,
Nine fawns, a calf, three skunks, wait there's more -
An owl, two sheep, two goats and some cubs
to feed and burp, with bellies to rub.
And Yankee, sweet Yankee, a mother's lament
my lion cub's gone, my heart in a wrench.
(That was the hardest part and I fear -
I must do it again, and again next year).
There's been meetings to attend, PR to do
Wow, we are always so busy here at the zoo.
Speeches and goof ups with egg on my face,
Like (surprise!) there is a dead snake in my case.
But chin up, my hawk, why he flew away
Gee folks, it must be rehab release day!
It's all fun, keeps me hopping, and damn
I'm always tap dancing as fast as I can.
We've almost settled into our new house
our cat is thriving, she caught a mouse.
Our tree is not up yet, but the yuletide decor
is warm and cheery as you come through the door.
Jack's doing well, in school she's straight A,
she's made lots of friends who come over to play.
Painted nails, fancy hair, she's quite the treat
a preteen daughter is really quite neat (?)
Giggles and screams. And then there's this guy,

THE ZOOKEEPER'S DAUGHTER

(can't mention his name or she would just die!)
Doug loves his work here on the farm.
(though he's become somewhat redneck - I'll admit alarm)
He helps pick up road kill to feed the big cats
he was grossed out at first, but now it's old hat.
Hay bales, grain bags, backhoe and tractor
have definitely increased the muscle factor.
Doug says he's always in blood, guts and s--t,
he's a number one guy and that's about it.
So to all my friends both far and near,
come up to see us this coming year.
I miss you all, wish you would write-
Merry Christmas to all and to all a good night.
Love, Lori, Jack and Doug

Chapter 39
The Lion's Tooth

On the March routine visual inspection of the pride of African Atlas lions at the zoo, I noticed a slight facial swelling under the left eye of Moses II, our 450-pound male lion. Moses II was the son of Moses, the cub I had raised when I was eighteen. I alerted Dad and Parker to the possible problem and the big cat was observed for the next few days. When the lion's facial swelling and behavior did not seem to improve, Dr. Ted Spinks, VMD of the Animal Hospital of Sussex County was called in. Dr. Spinks prescribed a regimen of antibiotics and scheduled a dental examination for the male lion. "Dental problems are fairly common in other zoos where a pre-packaged food product is consumed," he explained to us. This was our first experience with a tooth problem. Dr. Spinks was surprised. "The natural diet fed to the large carnivores here, helps keep their teeth clean and cavity free," he added.

A dental examination for a lion requires a coordinated effort of planning and scheduling. Dr. Spinks consulted with specialists Dr. Joseph McClure, an endodontist and a professor at Temple University, and Dr. Carl Tinkelman, an endodontist, who has worked with the Philadelphia Zoo for 25 years. A date was set for the examination after the lion had completed his course of antibiotics.

Anxious anticipation was apparent as the date drew near. "The health and safety of both humans and the lion are foremost in my mind," Dad mused to me. "Moses II is the third generation of African Atlas lions here. He has a gentle personality, but don't let that fool you, he is a lion by nature, and right now he has a sore tooth, which makes him more dangerous than usual." Dad was worried. We all knew that a hurting animal was more dangerous than usual.

Dad designed a special chute in the lions' den and Parker fed the lion his medicated food in the chute for a week. This was done to get the large cat accustomed to walking in and out of the confined area where the injection of anesthesia would be given. "We want Mosses II to wake up in his den, in familiar surroundings so he won't be upset. By setting up in his den we lessen the stress factor and the time he needs to be under anesthesia, by avoiding transport time," Dad explained. I am constantly amazed by my dad's animal knowledge.

The morning of the exam was cold, as would be expected of March weather. Plans were made to bring in heaters and generators needed for the lights and other equipment. The propane heaters had to be kept away from the hay and the oxygen containers. We all kept an eye on that, as we did not want to be blown to "Kingdom Come!" An anesthetized animal cannot maintain its body temperature and can easily catch pneumonia, so the heat was absolutely necessary.

Dr. Spinks, Dr. Servideo, and Dr. Bullock, all VMDs from the Animal Hospital of Sussex County arrived with Dr. McClure. The veterinarians had consulted with the University of Pennsylvania's Veterinary Anesthesiologist, Dr. Lin Klein, and an expert with the anesthesia of big cats. Dr. Tinkelman was on his way and the decision was made to start the anesthesia.

The females of the pride were locked out of the den. Parker lured the male lion into the chute, and the chute door was shut. Dr. Spinks administered the first injection of anesthetic by use of an extension syringe. An extension syringe is a needle on the end of a five-foot pole. Lions will not stand still for an injection. The injection was given intramuscularly and we all waited for the lion to lay down. A second injection was given and a few minutes later the sleepy cat dosed off.

The moments that followed next can only be described as an intricate step by step dance. Parker and Dad dismantled the temporary chute, handing out the individual parts to Doug and I outside the den. The external den gate was removed. Tables were set up inside the den. The huge lion was placed on a stretcher and six men lifted the 450-pound feline onto the table. Next, the portable anesthesia machine was lifted into the den. Dr. Servideo started the gas anesthesia. Dr. Bullock monitored the animal's pulse and respiration. Dr. Tinkelman arrived and assisted Dr. Spinks with the positioning of the mouth and tongue of the lion. Dr. Tinkelman brought with him special braces to keep the lion's mouth safely open. He had designed the braces for use on the big cats in Philadelphia. Dr. Spinks placed an intravenous catheter in the lion's foreleg for the administration of IV fluids during the procedure.

Family friend and documentary filmmaker Vic Campbell, of Buzz Creek productions, crawled into a "bird's eye view" position on top of the lions' den. Veterinary technicians from the Animal Hospital assisted the four doctors as everyone peered into the huge lion's mouth. Moses II slept as we discovered that the left carnassial tooth had abscessed and became loose. The decision was made to extract the tooth rather than perform a root canal, because the tooth could not be salvaged. A specimen was taken for biopsy. During the entire procedure Dr. Servideo and Dr. Bullock monitored the level anesthesia, pulse and respiration to insure the lion's and the humans' safety. Dr. Spinks then administered routine immunity vaccinations. This part of the procedure was over.

The dance was then reversed; the lion was unhooked from the anesthesia machine, equipment and people removed from the den, and the lion placed back on a fresh bed of hay. The gate to the den was hung back on its hinges. Dr. Servideo administered a pain suppressant and an injection of reversal anesthesia along with antibiotics. Needles were flying fast and furiously. The procedure that had taken two weeks to coordinate and plan was over in two and a half hours on a chilly March morning. The veterinarians and endodontists packed up their equipment. The family waited for the first signs of recovery. Dr. Spinks congratulated his team on a job well done and assured us that the lion was doing well and would wake up completely in a couple of hours.

Moses II began to move at two in the afternoon and was checked on every hour. At four p.m. the lions' den was opened and the rest of the pride (three females) were let in to 'comfort' the groggy Mosses II. At six thirty that evening, Dad called me with the good news: "Moses II was standing up and has moved to a fluffier section of the hay. He seems to be doing fine." We all slept a little better after that.

A week later the biopsy report on the sample taken from the lion's mouth came back. The diagnosis was mouth/bone cancer. The entire family was saddened by the report, but hopeful when Dr. Spinks said, "The Big Guy could have a day or ten years, depending upon how virulent the cancer is." The next couple of days Moses II gradually improved. He was seen frolicking with his pride in the large grassy run. The Vets and the family were hopeful. One week later, however, Moses II was again lying in his freshly hayed den, his face painfully swollen. He was in obvious discomfort. Moses II refused to eat. "You could just look at his face and tell he was suffering, his eyes told the story," Dad said with sadness in his own eyes. Dr. Spinks was called again. We gathered to pay our last respects. Moses II was euthanized.

Space Farms had three females in the pride and it appeared the two young females were pregnant. The family was hopeful that Moses and Moses II's family tree would continue with a fourth generation born at Space Farms. I crossed my fingers and all my toes.

Chapter 40
Blizzard and Cousins

Space Farms is located in the hamlet of Beemerville. Beemerville is tucked snuggly into the base of the Blue Ridge of the Kittatinny Mountain range of the Appalachians. Beemerville is in Sussex County, the last enclave of God's country in the northwestern part of New Jersey. Our little hamlet is surrounded by farmland. Not all of New Jersey has an exit on the turnpike. Because our little hamlet of three stop signs (no traffic lights, gas stations, grocery stores, or bars) is "way up north" at the foot of a mountain, we get a lot of snow on occasion. Beemerville is usually fifteen degrees cooler than New York City. Often the City gets a dusting of snow and we get a foot. Snow can happen here from October to April. The spring of '97 gave us that kind of white surprise.

On April 1st the flurries were coming fast and furious. The next morning we awoke to a frozen deposit from a major blizzard. The men were out early on tractors and in trucks to plough the snow away. My car was snowed into our garage; we plough out the animals first, as they need feed and water more than I need a car. Makes perfect sense to me. Jackie was excited about all the snow. (What kid isn't?) I told her to stay in the house and I was walking down to the zoo. I trudged through the snow to the zoo (not a big deal, it's only a block away), and helped to shovel walkways and doorways to the main building. Then I put on coffee, knowing the men would be cold after all the snowy work they were doing.

My brother Parker was the first to slosh into the kitchen from the zoo. I noticed his heavy coat was unzipped part way. He was blowing on his fingers to warm them up faster. After he brushed himself off, he turned to me and said, "Here, I brought you a present, but he's a little cold." Parker put his huge, cold, calloused hand inside his warm coat and pulled out a small Atlas African Lion cub. My six foot bundled brother had seen the little lion lying in a snowdrift outside the lions' enclosure. "His Mom must not of wanted him." Parker shrugged and went about the process of coffee and warming up.

The cub was good sized, two and a half pounds, covered in soggy tan fur with dark brown spots just like all the cubs before him. Parker had saved the cub's life by tucking him inside his jacket, letting the little lion Popsicle absorb Parker's body heat. When I first held the lion, his tiny tongue, paws and ears were still cold. This baby had fortitude, and the good fortune to be found by my brother in a sea of snowdrifts.

I immediately hurried the cub off to the office, dried him off with a towel, tied off and trimmed his still wet, partially frozen umbilical cord. I set the cub up on a heating pad, and walked back to the kitchen to mix formula. A warm body and warm formula in his tummy would settle his immediate needs. I worried about frostbite on his ears, paws and tail, but only time would tell. *When would this mother start taking care of her own babies? Why didn't she keep him in the warm, dry, hay stuffed den?* I wondered. But in truth, I didn't mind at all.

As soon as the formula was mixed I loaded a bottle with two tablespoons and a soft beginner nipple. I trimmed my index fingernail to the quick, so I wouldn't scratch his mouth (my work does not allow any fem fatale frivolities). I gently pried cold lion lips open to insert the nipple. He sucked the bottle dry in two seconds flat. I am always amazed at the differences between babies. Some drink like there is no tomorrow; some lollygag on the bottle while watching the birds and clouds overhead. I stimulated him for urine and let him sleep. He was content, temporarily, and settled down motionlessly on his heating pad.

I headed for Gramma's house to tell her the good news. Halfway up the path to Gramma's, Dad was still shoveling snow.

"How's the cub?" he asked.

"Fine, Pop," I replied, "or at least as well as can be expected."

Dad finished the last few feet of snow removal as he reminded me, "You'll have to keep an eye on him for frostbite and pneumonia, those little lungs freeze up fast."

Don't remind me, like I wasn't already worried enough, I thought. We went into Gramma's to see how she was doing and get some hot coffee. Dad and I told Gramma the good news about the lion cub and I filled her in on the care I had given him.

"You gotta watch him for pneumonia and frostbite," Gramma warned me. "Let me know if you run into trouble."

Gramma knew so many tricks of the trade, things you'll never find in a vet manual, and her suggestions usually worked. Then Gramma's worried look dissipated, her old eyes twinkled, and she said in her perky, "I've got a good one for you" voice: "You'll have to call him Blizzard." And Blizzard it was.

Blizzard did very well, guzzling his bottles down in record time. He put on weight eating every two hours. He showed no signs of permanent damage from his time in the snow bank. Blizzard was a lover. He enjoyed the cuddling that Doug, Jackie and I lavished on him. An only child, and the only baby in the nursery at the time, he was destined to be spoiled. Oh, OK, I'll admit it, I spoil most of my babies. Blizzard developed on lion schedule and was increasing in size in proportion to the increased allotment of formula he was taking in. That lion milk must be great stuff, nutritionally, but it tastes like chalk to me. I've learned the hard way to keep all animal infant formulas labeled in my refrigerator.

The sweet, dark spotted cub traveled with me to work in the daytime and home at night just like the other babies I have raised. Separation anxiety affects animal infants, too. Even though I am the adopted mom, the babies want to be with me—and I with them.

Blizzard was three weeks old, toddling around the house when my world got crazy. One O'clock, the mother of Yankee and Blizzard, was named for the freckle on her nose, a black speck on a pink nose. Three O'clock is the larger, heavier mom that had the original three cubs I lost. Both Moms were born and raised by their own parents in the same enclosure they were now in. On April 25th, Three O'clock had her litter. It was Three O'clock's second litter, and she was not paying enough attention to the young. Dad decided we would bring the litter in. I would like to reiterate that 80 percent of lion cubs born in the wild do not make it to their second birth day, and on top of that, it usually takes a new mom a couple of litters to get the idea of what to do. I hoped the moms would do better next time, but I'll admit I was not adverse to the task at hand.

We brought the five lion cubs into the office/nursery. I recruited Doc and Barb, the kitchen supervisors, their visiting granddaughter, Jenna and Alex, my cousin's son who was working at the zoo for the summer, for an emergency nursery crew. The five of us each worked on the cubs. Each cub was swaddled in a clean towel and given a brisk rubdown. This both stimulated the cubs and helped clean off the birth fluids that mom lion had neglected to do. Doc and Barb were grandparents and knew exactly what to do. They helped supervise their granddaughter. It was really sweet to watch Alex, who was single, six foot four, and twenty-six years old at the time, enjoy his new emergency fatherhood status. Everyone thought their cub was the best. Alex was particularly proud; he had the only male cub out of the litter of five. Sometimes the job is yucky, but the joy is worth it. I worked on all the cubs checking umbilical cords and for other potential problems. Two cubs had minor puncture wounds that I thought would heal on their own, but I would need to keep an eye on.

I weighed the cubs and set them all up in individual boxes. Young infants constantly want to suck. If the litter was left together, the cubs would suck on each other and drink whatever fluid they came across. Needless to say, this would make them sick. I requested heating pads from neighbors, from friends, and got two from Gramma. Each cub had its own cardboard box, towel, snuggle stuffed animal toy and a heating pad. We left the cubs set up in the office to settle down while I started to mix more formula. All the while, Blizzard watched, I'm sure wondering what his new status would be with all these new babies. What a job I had ahead of me.

I weighed the cubs; Alex's male was the tiniest, under a pound. The four girls varied from a pound and a quarter to two pounds. Blizzard at three weeks was eight pounds. The larger the litter, the smaller the weight of the babies. The girls were strong and good drinkers when offered their first feeding of two tablespoons. Alex's male was a different story.

In order to keep records straight, monitoring formula intake, poop and urine output on the six cubs, I color coded the bottle rings, handmade small collars, and their names. Sky was blue and the only male. Dandy Lion was yellow, Violet of course was purple, Rose was red and Petunia was pink. Sounds hokey but what is the little old lady who lived in a shoe to do?

After the cubs were born, I called the papers to announce the birth of the rare Atlas African Lions. Reporters and photographers came from all over to take pictures. It was not easy trying to hold six squirmy lion cubs at one time. The photographers had a good laugh with me, as the cubs tried to crawl, wiggle and squirm, with their eyes closed. Blizzard was a trooper, and he cuddled close to me, watching the pandemonium.

The next week passed in a blur of fast forward perpetual motion. I had six boxes in the office, six cardboard boxes in my car and six boxes in my house. It was easier to carry just the cubs than the boxes with heating pads and towels. Towel laundry was done twice a day, boxes being changed if the babies made an accidental mess. The cubs were fed every two hours around the clock for the first week. It took one hour to fix the formula, feed and potty each cub, and make sure the boxes were clean. I was exhausted. And Sky was failing. He never drank well and seemed lethargic most of the time. I called Dr. Servideo who suggested we start him on IV fluids. Sky declined and I lost him during the night. He was less than a week old.

Every morning Dad would see me pull in with my car full of lion cubs and from wherever he was he would give me the six fingers up question. I would give him the six fingers up with a smile. It was a hand signal code we had developed over the years. Each finger represented one baby. The morning after Sky's death I gave Dad the five fingers up and a shrug. Dad knew the tiniest cub had been failing and had died. The official diagnosis was failure to thrive. It is common in nature for large litters to lose the runts. I know this in my head; my heart is another story.

I settled the cubs into the office nursery and sulked up the path to Gramma's house. Gramma consoled me, as only she could do. She had the experience and the knowledge that I respected enough to accept her viewpoint on the situation. When I started to tear up, she quickly snapped at me, "You've still got five babies, don't fall apart now, they need you." Gramma came down to the main building almost every day for lunch. When she came down that day she saw the other five cubs and pronounced them healthy. Hearing that from Gramma made me feel more confident than if I'd have heard it from a vet. The girls and Blizzard were doing great; I didn't need to worry so much. At three weeks old two of the girls, Rose and Violet developed bladder infections. I called Dr. Servideo. Ten days on an oral antibiotic and they cleared up fine.

The five cubs were quite the attraction at the zoo. I was constantly asked questions about them. I developed a speech and gave it every day at one feeding, outside on one of the picnic tables in the zoo. The visitors loved it. I would often have a hundred people standing around listening to me talk about the cubs and demonstrating their care. The people were kept back from the table and not allowed to touch the cubs (people germs, you know). I even demonstrated the poopy part, stimulating for urine and feces. Explaining the job is never complete until the paperwork was done. Adults laughed knowingly, kids just said, "Euuwy!" I was proud of my babies; I loved to show them off. I must admit I was also proud that crowds of people were listening to me just like they listened to Dad's snake lecture. Dad and I would often joke about who had the biggest crowd.

Space Farms had made an animal trade and we owed one lady a female lion cub. I had kept a list of zoo people who wanted lion cubs after putting the ad in the zoo trade magazine for Yankee. When I felt the cubs were strong enough and able to go without a bottle for eight hours I started the hardest part of my job. I had to place the girls and Blizzard. I did phone interviews of the potential new keepers, zoos and sanctuaries that had requested lion cubs. I turned down a couple of people; I'm very picky about where my babies go. There are a number of federal and state licenses that are required for an individual, zoo or sanctuary to obtain a big cat. I require all of those before I will talk to any prospective new parent. And then I want to hear certain questions that let me know that the person on the other end of the phone knows what they are doing. If they pass my test, then I will consider them for ownership. I've turned down celebrities. Money can't buy everything.

Dad started to build boxes. USDA shipping requirements are such that all big cats, no matter the age (or the fact that my little babies could not fight their way out of a paper bag) are shipped in plywood and steel boxes. Dad had been building animal boxes for fifty years; he knew what he was doing. Ventilation, watering access, and security were all part of the process of designing a box.

Dandy Lion went to Nevada, on an airplane. Her new mom had big cat experience. The trip to the airport is the worst drive for me. The traffic to Newark airport is always a white-knuckle drive. Throughout the hour and a half drive I have visions of potential accidents, of me being injured and a helpless lion cub running petrified through four lanes of fast moving cars with the inevitable results. I don't like the drive. After I arrive at the airport, the paperwork is intensive and a security check of the box completes the official process. Then I have to leave my precious baby in the noisy cargo hanger. As I leave I tell myself not to look back... don't look... don't look. I always look. The last vision I have is of my special cargo peering out the wire window with a "Mommy, where are you going?" question in her eyes. Not panicked or upset with the box. Dandy played in hers for days to get used to the box. Just, "Where are you going and why can't I come?" I knew Dandy was fine, her tummy was full and she would sleep soon. But I looked. I saw the question I could not bear to answer and headed for my car. Inside the car, my head resting on the steering wheel, I sobbed. I thought, *I have to do this again and again; I hope it gets easier.* It doesn't. I dried my eyes and started the white-knuckle drive home. When I get home, the other cubs are waiting for me, with perky eyes and wagging tails. I played with them a little longer at bottle time and stayed close to the phone. The emotional shadow of the drive, the farewell, and the long, lonely ride home hung over me for the day. I must give credit here to my family and friends who work with me.

They know how hard it is for me when my babies have to go and cut me a lot of slack in the rest of my workday. I'm sure visitors to the zoo wondered why the zoologist is walking around red faced on the verge of tears, but I managed to avoid making eye contact.

I had left Dandy at the airport cargo bay at eight in the morning, got home about ten, took care of the cubs and waited by the phone. Every phone call, and the zoo gets hundreds of phone calls a day, I jumped and looked toward admissions where the phone was answered. A negative shake of the head tells me the call is not for me. I heard from Dandy's new mom at ten that night. She called me at home to let me know that Dandy arrived safe and sound. Dandy's new mom was thrilled with the condition of the cub, and Dandy's personality. I fed the other cubs, went to bed and slept like a log. I was up at six the next morning to start my cub feedings again.

Rose and Petunia went to Michigan. Their departure was easier on me. Their new dad was the owner of a trucking company who ran a sanctuary for animals. He wanted two females to round out his collection. He came with an eighteen wheeler and spent the afternoon with me in the zoo. By the time he drove away with the cubs in the cab up front, I was assured that I had made the right decision for my girls.

I had a potential dad for Blizzard but was not sold on the idea of Blizzard going to a private owner with little experience himself, even though his family had a circus animal background. So I stalled, not returning phone calls, avoiding the situation. One day, Blizzard's potential dad showed up with blue prints of a wonderful enclosure he had built with big cats in mind. That did it for me. I knew Blizzard would go to a wonderful home, with someone who would give my big boy the love he deserved. And Blizzard would be in New Jersey, close enough to visit. I gave my blessing to the placement, but Blizzard would have to stay at the zoo until some paperwork was cleared up. Gee, that was too bad! Blizzard left my care at twelve weeks old, one of the largest twelve-week-old lions I had ever raised. At twelve weeks he was twenty-four pounds of cuddle, a big baby, a lap lion.

On Blizzard's last day with me I was to take him to our vet to be declawed. I tried to talk his new dad out of it but he was insistent. Blizzard's new dad was to pick him up after work. I held Blizzard in my arms while Dr. Spinks gave him an injection to put him to sleep. I told Dr. Spinks and Dr. Servideo to call me if there were any problems. The operation went fine. The problem was Blizzard was too big for their holding facilities and Blizzard's dad could not pick him up till five o'clock. After Blizzard woke up he was not content to stay in the holding pen sized for a regular large dog. He wanted to be out by the vets and vet techs. He was a people lion. I rushed to the vet's office fifteen minutes away. I walked in to find Blizzard stretched out on the floor, playing with one of the vet techs. All four feet were bandaged. I burst into tears. Dr. Servideo put her arm around me, said, "Come on, tough it up, girl," and assured me Blizzard was doing fine. I called to Blizzard and he pranced tenderly toward me. I scooped him up and hugged him while he licked my face. After I calmed down the vets helped me put Blizzard into his transport kennel for the trip home. Blizzard was fine, I was a wreck.

I brought Blizzard to my house and set him up on the front porch, which had linoleum flooring. With bandaged feet he could not go outside on grass or dirt. My daughter, Jackie, now eleven years old, was to stay home and be in charge of babysitting the somewhat groggy cat. Blizzard was lying on a quilt cat napping when I left to go to town for the groceries for the zoo restaurant. When I returned from town an hour later, there were frantic messages left for me by Jackie. I zoomed to my house, and was met by the back door by my sobbing daughter and my mom, Beverly. Jackie had called Gramma Beverly when Blizzard tore off one of his bandages and Jackie saw a little blood. I assured Jackie she did the right thing. I left Mom in charge of calming Jackie while I attended to Blizzard.

I called my friend Karen Martin who lived up the hill from us. Karen was a vet tech before she had kids, and I knew putting the bandages back on was going to be a two-person job. Karen and I replaced the bandages with fresh ones—by this time Blizzard had gotten another one off. We secured the bandages with duct tape. Duct tape is a wonderful multi purpose invention, and the only thing I know that keeps bandages on big cats.

And how was Blizzard you ask? Blizzard loved all the attention he was receiving, perched first on my lap and then on Karen's. Mom and Karen kept vigil over Blizzard while I went back to work for a few hours. I had other babies to take care of. Blizzard eventually tired out after his long day and fell asleep just in time for his new dad to pick him up and take him home.

I hate de-clawing. De-clawing is the surgical amputation of ten digits. Think about it. I know it is standard practice, but if you are going to deal with a big cat, you have to deal with the whole cat, not just the safe parts. I will never volunteer to babysit a recently de-clawed cub for anyone again.

By mid-summer I only had Violet left. And boy did I spoil her. Violet came home every night till she was ten weeks old. Her first night out was tough on me, but she did fine. She would come out for speeches on the picnic table like a pro. She loved the public and performed her "sit," "open," and "roll over" on command. Eating venison, rough housing with toys (I do not allow rough housing with me) and meowing to me filled her happy days. Somehow, I just didn't find a home for her till the end of the summer. Violet was placed with a trainer, and is working with other big cats.

Chapter 41
Jimmy the Llama in Pajamas

The camel family originated in North America, according to fossil records. The smallest were the size of rabbits, the largest fifteen feet at the shoulder. As the species grew and diversified, they spread across the Bering Straits land bridge into Asia, south to Africa and some went due south to South America. Many of the different species died out, leaving the two humped Bactrian camel (from which the one humped dromedary camel was domesticated) in Africa and the South American Llamas.

The South American llama lives in the high mountain plateaus of the Andes Mountains in South America, usually between 13,000 and 16,000 feet above sea level. There are three subspecies of the llama family, Llama vicungua, Llama gunaco, and Llama llama. Space Farms Zoo had Llama llamas, which are about the size of a small horse. Llamas are shy and possess exceptionally good eyesight and hearing. The eyes of the llama are huge rounded orbs with luxurious long lashes. The protruding eyeballs enable the llama to see one hundred and thirty degrees from the front center, an obvious advantage against predators. The long lashes protect the eyeball itself. The combination of the large eyes and long lashes give the llama an appealing, doe eyed, sweet and innocent appearance. Llamas come in colors of black, brown, white, chestnut, pumpkin, and/or any combination thereof. The abundant thick fur is an insulating factor in the winter and the summer, enabling the llama to endure extreme temperature variations. The fur is shed naturally, giving the llama a natural clumpy Rastafarian look if they are not brushed.

The main defense of a llama is to run, fast. Or if they think they can intimidate the foe, llamas will belly bump (like a sumo wrestler) kick or spit. The spit is actually chewed cud, the llama being a bovine. That fact does not make it any nicer if a llama spits on you. Green slimy spit does not accessorize any outfit well. We always know who is behaving in the zoo and who is teasing the animals. The visitors that tease the animals always come in to buy a clean T-shirt from the gift shop!

Llamas are herbivores, feeding on grasses and grains, browsing constantly during the day. Llamas are herd animals, a herd consisting of one male, a harem of females and young. When young male llamas mature they join a teenage male herd, jousting and practice fighting, waiting for their chance to fight for supremacy of a herd. Mother Nature has equipped the male llamas with slightly bucked bottom teeth. Llamas have no top teeth, only a hard palate. The head honcho (alpha) male of a herd will attempt to castrate the younger males that challenge his supremacy. It is not a pleasant sight. We always remove the young males before sexual maturity. The South American natives domestically breed llamas as pack animals. The natives use the llamas as pack animals, comb out their fur for yarn, milk them for dairy products, and on occasion, while stuck in an Andes snowstorm, eat them.

Cria (the Spanish word for llama babies) are born after an eleven-month gestation and walk within an hour of birth. By five hours old they can run with the herd. The neat thing about llamas (that we keepers and the vets appreciate) is that the cria are always born in the daytime. Dad explained it to me. The babies could be born year round but any baby born in the cold winter months would be doomed to freeze to death if it was born (and stayed wet from birth fluids) at night. Cria born in the daytime have the warm sun to help them dry off and get the thick coat ready to insulate them for warmth through the night. Here is an evolutionary survival trait of being born in the daytime. Makes sense if you think about it. Another interesting fact of llama husbandry is that llamas always pee and poop in one spot, making them easier to clean up after than, say, horses or cows, that leave patties all over the pasture. The alpha male will mark this territory with his own urine.

The Space Farms Zoo had purchased a pair of llamas in the early nineteen eighties. The herd increased naturally as time progressed. In the mid eighties the llamas became a fad species for gentleman farmers looking for unique animals to add to their farms. The natural territorialism of the males and females made them a safe alternative to guard dogs. Llamas are used to this day as protection for sheep herds and other herd animals. Llamas challenge intruders, belly bumping, kicking and spitting, but rarely doing any permanent harm, while intimidating their foe.

The price of a female llama in the late eighties was $10,000 for a newborn. The fancy purebred breeder llamas were selling for upward of $50,000. It was easy money for gentleman farmers. By the nineteen nineties the cost for a pet quality llama had dropped to $500. The fad had played itself out, a classic case of supply and demand, but the popularity of llamas had not.

Space Farms Zoo has a standing herd of fourteen females and one male. Every year the young males were sold off for their own protection from the alpha male of our herd. Young males were left with mama for a week or two, then taught to drink kid goat milk from a bottle and placed with a new owner. In the spring of 1997, a male cria was born. Cria are about the height and length of a calf, but only weigh twenty to twenty-five pounds. They are all fluff and bones. As per our procedure with new births, we observed the birth and the condition of the newborn. The young cria was having trouble walking, bearing no weight on his right front leg. The decision was made to bring the baby cria down to me in the nursery a few hours after birth instead of leaving the cria with mama for the next two weeks.

Dad, Parker, Doug, Keith and I went in the two-acre paddock after the cria. The entire herd was cornered in a section of the paddock, the alpha male approaching Parker. There is not a braver man whom I know of than my brother Parker is when it comes to facing down an animal. Parker has height and the sturdy girth of a linebacker, and animal behavior knowledge that enables him to predict an animal's next move. He can intimidate any animal he has worked with (and a lot of humans I know). The male challenged Parker, charging to belly bump. Parker out maneuvered him, sidestepping at the last moment. The rest of us let the females past us until the mama with the fuzzy fawn colored cria came past. Dad lunged for the cria and got him in a headlock. After the cria was captured a strategic shift took place; Parker grabbed the baby cria in a whole body hold, his arms around all four legs. He then lifted the twenty pounds of kicking fluff and walked quickly toward the gate. At this point the rest of us have one objective, to protect Parker (or whoever has the baby, this time it was Parker) from the alpha male and the mama. The rest of the herd had little or no interest in the abduction at this point and ambled away. Dad, Doug, Keith and I looked like guards on a basketball team, hands in the air, surrounding Parker and the cria. Hands high in the air to make ourselves look larger and more intimidating to the male and female, we escorted Parker and the cria to the gate. The female was not capable of understanding we wanted to help her injured baby, and the male was simply responding to the intrusion of his territory.

Once outside the gate, I climbed into the back of a jitney, sat on a milk crate, and received the little llama. Parker and Doug drove me to the nursery.

Llamas cush. Cushing is a complete collapse of the legs when under duress; in other words they give up easily once pinned. The little llama cria had stopped kicking and rode gently to the nursery. Dad and I examined the leg for a break or other reason for the limp. We could not see one. Dad said that often dairy calves would be born with a limp due to their position in the womb or a birthing trauma. Birthing traumas are common in hoof stocks that give birth in a standing position. Hoof stock is usually born in a diving position, with the two front feet on either side of the head. Dad manipulated the little llama's legs and discovered the right front leg slipped easily over the head, to the left. The muscles had been stretched out during birth, the baby being born in a lopsided position and then falling to the ground. That had most likely caused the strain on the leg and the limp. Dad had seen many dairy calves recover on their own after this kind of birth trauma with enough exercise.

I let the cotton soft, light brown baby roam free inside the nursery fencing for maximum exercise. I crossed my fingers and waited for him to get hungry. I named him Jimmy, after a talented sculpture artist friend of the family who was born with spinal bifida and also walked with a limp.

Training a llama cria to drink from a bottle is no easy task. Newborns are easier to break to a bottle than a cria that has been with mama for a couple of weeks. Nipple confusion happens to animals, too. Of course the baby is frightened of you, you are a strange creature to them. They can run fast. I leave them alone for about six hours to get used to their new environment; even chickens can be scary if you've never seen them before. After being away from mama for about six hours they get hungry. That's when I first try them on a bottle. We use the kid goat milk replacer and a standard human baby bottle with an elongated nipple. Llamas cush and I use this natural phenomenon while training the infant. I slowly approach the baby and wait till the last possible moment to reach out and grab a hold of them. There is no use running or trying to chase young llamas down, they run faster than I. I use both hands to settle him down and swing one leg over its back. I seem to be just the right height so I am not riding them with my body weight, but gently corralling the cria between my legs. One hand rubs the neck while the other hand slowly comes up with the bottle between the thumb and forefinger, nipple toward the palm. Using the middle or ring finger of the bottle hand I ever so gently pry open the jaw behind the front teeth (cria are born with teeth) and insert the nipple. If the cria does not suck right away, I rub the neck area. I often cover the baby's eyes to eliminate distractions. If they were drinking from mama, the eyes would be closed, chin resting between mama's back legs. As soon as they figure out that the milk flows when they suck, they drink greedily. I remain as quiet as possible so as not to scare the cria with a strange voice.

I learned a trick from a llama herder, Cathy the Llama Lady. I gently blow on the nose of the llama baby to "scent" him. The mama llama would nuzzle and breathe on her baby while nursing. This helps with the bonding process between the cria and myself. So here I am straddling a tiny llama, bending close to shade his eyes with my chest and arms, blowing on the nose, all the while holding a bottle in reverse position with one hand, rubbing the little llama's neck with the other. I often look like a contortionist while feeding llama babies. To add to all of this maneuvering, I had to be careful of Jimmy's sore front leg. Visitors to the zoo nursery often wonder what in the world I am doing, shouting questions to me during this process. I'm sure they consider me rude when I do not answer them. If I shout back, the baby would be upset. So I simply do not answer, concentrating on the task at hand. Afterward I explain the process and apologize for not speaking to the visitor till I'm done. Everyone seems to understand.

The entire procedure takes about ten minutes in the beginning. After two or three days of six times a day, the cria settle down, realizing that I will cause no harm and I have the answer to their hunger. By day four the cria usually come running to me for a bottle and put their prehensile, rubbery, cleft lips on the nipple without trouble. It just takes time, patience, and a very calm demeanor.

Jimmy broke to the bottle easily. Within two days he would come limping over to me, nuzzling, looking for the bottle. By day three he was putting his own lips on the bottle and drinking freely. His limp was slowly improving, his shoulder no longer tender to the touch. After Jimmy bonded to me and accepted me as his mom, I took Jimmy for walks around the zoo, outside the nursery fencing. This was good exercise for his leg. OK, it was good exercise for me, too. Jimmy would follow me around just as he would have been following his herd. If something scared him, he would stick to my side like glue. When he felt comfortable, he would trot away from me and then come running as fast as he could back to me. I cannot put into words the joy I felt the day I saw him wander away and come running back to me kicking his hind legs high in the air. It had taken about three weeks until his front shoulder was able to support the weight of his body. He galloped and frolicked with all the elation only a youngster can feel.

Jimmy stayed in the nursery for the summer. One day, as I went out for the eight a.m. feeding, I noticed Jimmy's side was dripping blood. I corralled him once again between my legs and searched his thick fur for the wound. Somehow Jimmy had received a three-inch slice through the skin on his ribcage. The thick fur had covered the wound so I did not see it until it was coated in blood. I called Dad in. We gave Jimmy some emergency first aid cleaning out the wound and bandaged it tightly against his ribs. Dad inspected the nursery fencing to find out what had happened. We still don't know how Jimmy got such a bad gash on his side. There were no tufts of fur stuck on the fence or sharp objects in the nursery. We kept an eye on his wound. When it did not appear to be improving, I called Dr. Servideo. Dr. Servideo would see him as soon as I could get Jimmy there. I pushed Jimmy into the back of my SUV and headed down the road. Jimmy settled down quickly, sitting with his long neck and head looking out the window. You can imagine the strange looks we received at every stop sign and red light.

Due to the slipperiness of the linoleum in the veterinarian's office, Dr. Servideo decided to work on Jimmy outside on the grass, so Jimmy could stand. Under a tree, Dr. Servideo shaved the fur off the site of the injury and cleaned the wound thoroughly. Jimmy was a trooper, never making a sound, corralled between my legs. I packed Jimmy back up into the car and brought him home to the nursery. Dr. Sevideo gave me medications and instructions for the daily care of the wound.

Jimmy was happy to be home and the first thing he did was run to his sand pile to dust. Llamas, horses and buffalo love to dust in the dirt. I was flabbergasted. This would not be good for his wound with all the goopy medicine on it under the bandage. If dirt got inside, it would cause an infection and more problems. *What to do? What to do?* I had a brainstorm and rushed home to my house. I grabbed one of my daughter's old turtlenecks. It did have an animal print on it, but that was secondary to the fact that the turtleneck was the proportion that I needed. I called my friend Karen Martin, asking her to meet me at the nursery. When Karen arrived, we cleaned the wound's bandage of dust and dressed Jimmy in the animal print turtleneck. Llama cria have very long legs. The neck fit jimmy well, and the arms of the turtleneck reached to his mid fore legs. However it was not easy getting those long legs into the arms. After much ado we accomplished our goal. The torso of the turtleneck was much too large for the little llama so we tied a very fashionable knot on his back.

We were standing back admiring our work, when a visiting family came through the zoo's entrance door next to the nursery. Their five-year-old daughter giggled and exclaimed, "Look, Mommy, a llama in pajamas!" Karen and I laughed heartily, too, a good laugh after a hard job well done. I must admit, the beige baby llama did look funny in leopard print pajamas.

Karen stopped in daily to help me with Jimmy's medications and new bandage. Jimmy would drape himself over Karen's lap while she was sitting in a chair. I'm not sure if he was cushing or just knew that what we were doing was good for him. He dusted daily, needing new turtleneck pajamas. He had quite a selection to wear, as my daughter would not wear the turtlenecks even after I washed them. Kids… go figure. Within two weeks the wound healed enough to remove his "pajamas." The fur was growing in and no scar would be visible. Jimmy was fine, and had become one of the friendliest animals I ever raised. Jimmy stayed in the nursery till fall and eventually went to the same home as Blizzard the lion cub. Jimmy is now an adult. I see his new dad every now and then, and he keeps me posted on his well being.

Chapter 42
Gramma's Snowflake.

Screech owls are the second smallest regularly occurring type of owl in this section of New Jersey. An adult screech owl is only eight to ten inches tall but can still rotate its head 270 degrees, like its larger cousins. The feather tufts on the top of the head give the appearance of ears, but like other birds, there is no external ear cartilage. The tufts are used for communication. If a screech owl feels threatened, it will stand up as tall as possible and extend the feather tufts to make the diminutive owl appear bigger to intimidate its protagonist. The actual ears of an owl are located on either side of the proportionately large skull at forty-five-degree angles below the eye socket. In most owl species the ear structures are large and slightly asymmetrical. One ear structure is situated higher than the other on the skull is. This helps with the sound location of prey. Owls are known for their excellent hearing. They are able to hear prey and then move their heads to triangulate the location of that prey.

Screech owls come in three colors: gray (like Snowflake), brown, and red, the three colors often showing up in the same brood. The large yellow eyes see very well at night; most owls are nocturnal. Screech owls have many vocalizations, but have been named for their screaming calls. The screeches of a screech owl sound more like the scream of a woman, and have often been mistaken for the scream of a bobcat. People find it hard to believe that such an ungodly sound could come from such a tiny bird.

Screech owls breed in March or early April, nesting inside hollow trees, abandoned buildings, or other birds' abandoned nests. After a 28-day gestation, owlets hatch from pure, white eggs. Screech owls mature incredibly fast. By four weeks old they are ready to leave the nest, but do stay together in a family unit, while mom instructs the young in hunting skills.

In April of 1984, a neighbor was cutting firewood. His power saw cut into a screech owl's nest. The mother flew away, the tree was felled, and the nest site destroyed. The neighbor, realizing what had happened and that the mom owl would not come back, brought a quarter sized, white, fluffy, hatchling owlet to my Gramma to raise. Gramma loved the little owl, naming it Snowflake. Snowflake was fed every two hours with little bits of meat on a toothpick. The little owl grew up and became quite attached to Gramma. Gramma would spend part of every day taking care of the owl, talking to it, stroking it, and cleaning Snowflake's cage. As Snowflake grew, Gramma increased its diet to include all types of meat, roughage (small bits of fur, sand, twigs, etc), and to my disbelief, lettuce or greens.

In the wild, the diet of the screech owl includes a variety of small rodents, insects, and often birds as large as they are. In none of my college zoology books were fresh greens ever mentioned in the diet of birds of prey. I assumed Gramma fed the owlet greens on the same basis that Gramma had always insisted that we humans have two colors of vegetables on our table. The family often joked about eating enough vegetables to live till tomorrow. Anyway, Gramma fed the owl greens on a daily basis. For years we thought Gramma's insistence of feeding her owl greens was part of Gram's advancing age. It was not until I was raising the three Kestrels, letting them fly free, that I realized the birds were seeking a nutritional need from green grasses and grains. Gramma was a lot smarter than I gave her credit for. I have always included greens in the diet of birds of prey since then.

In the spring of 1997 Snowflake was hale and hearty, enjoying watching zoo visitors from his perch inside the main building at the zoo. Snowflake would sit as still as the stuffed animals on the wall. Then, when an unsuspecting visitor stopped to look at him, Snowflake would turn his head or open one eye. This of course would cause one of two reactions—laughter or screams.

Snowflake was thirteen years old and was doing fine. Gramma was not. Gramma had started to fail, her age finally catching up to her at 93 years old. However, she often came down to the main building for her favorite lunch, a hot dog and French fries. That summer Gramma skipped some lunches, and I would take care of Snowflake.

I had seen Gramma work with Snowflake a thousand times, opening the door, petting his head, cleaning his papers, giving him fresh meat, greens and water. Gramma would click her false teeth to talk to her owl. Snowflake would bob his head and "converse" with Gramma with the same clicking sound. There were multiple layers of communication going on between Gramma's coke bottle glasses and the wide-eyed owl. Snowflake did not behave for me the first couple of times I was to care for him. When I told

Gramma about it, I felt like I was tattling. Gramma informed me that I was caring for him out of order. After I assumed Gramma's routine, Snowflake handled the cleaning much better, but would not let Dad or myself touch him.

Gramma had pneumonia, and had to go to the hospital later that spring. Snowflake stopped eating. Gramma recovered, came home, visited Snowflake, and Snowflake ate again. Later, Gramma would go to the hospital again and Snowflake died in my care. The average life span of a screech owl is six years. Snowflake was thirteen years old. A testament to Gramma's love and care. It was decided not to tell Gramma that Snowflake had died until she came home from the hospital.

We gave Snowflake's body to a local taxidermist to have him stuffed for Gramma. I went to visit Gramma at the hospital, and somehow Gramma already knew that Snowflake had died. Gramma's large coke bottle glasses steamed up, her gnarly, arthritic hand covered mine and she said, "It's ok, honey, he was old, like me." Gramma came home from the hospital, but was not doing well. Her life was ebbing away, we all knew it, and it was time. She had lived a full life for ninety-three years, surrounded by her family. You can say all the platitudes you want with your mouth, but when the time comes your heart still aches.

Chapter 43
Solomon, a New King for our Pride

After the death of Moses II, we were on the lookout for a new male Atlas African Lion. Dad put the word out that we were looking with all his animal contacts. I put an ad in the zoo newspaper titled, "Lionesses Lonely Hearts Cub In Need of New King for Pride." We received a few phone calls on adult lions that were extras and needed new homes. Dad was waiting for an Atlas or Black Mane African Lion. Space Farms Zoo had black mane genes and we did not want to dilute the strain. Another zoo in New Jersey was going out of business and we received a call—were we interested in a black mane older lion? Dad and Parker discussed it and decided to say yes. "Solomon" was supposedly twelve years old.

Solomon was brought to the zoo in July. He was old, a little on the thin side and road weary. He looked a sad sight, a few bumps and bruises from the trip, his eyes showing that his spirits were shook up. It is standard procedure to keep new animals in quarantine. Dr. Spinks came to do prophylactic blood work just to make sure the lion did not have any diseases that would infect the rest of the pride. When the blood work came back clean, Solomon was transferred onto the zoo grounds, to the concrete section of the lion's enclosure. The ladies were locked out on the grassy section. This was to prevent territorial fighting and to introduce the ladies slowly to the new male. We all observed the threesome carefully to see if the females would accept the new male. We had been told that Solomon had never been with females. It would be interesting. We worried a little, Solomon had been de-clawed and all the lionesses had teeth and all their claws. Lion fights are not easy to break up.

After two weeks Dad decided it was the big day. We waited till the zoo closed; just incase fireworks erupted, we needed to be prepared for anything and did not want to deal with the curious public. You never can tell what could go wrong, and catfights, especially big catfights, are not pretty.

The slide gate closing off the females in the grassy section was opened after the three cats had been fed. The females rushed into the concrete section straight toward Solomon. Solomon backed into the corner as far as he could go. Not a sound was heard from the lions or the humans as we all held our breath. The ladies started the welcoming with sniffs, and barred teeth, a lioness smile. Quickly the welcoming progressed into rubbing, licking, and courting meows. The ladies liked him. Halleluiah! Solomon was not on the same page as the ladies. He had never seen another lion before, or even been close to one, let alone two females in their sexual prime. Solomon backed up as far as he could into the corner, and was at one point up on his hind legs, front paws swatting the air. One of the ladies crouched on all fours and backed up to the petrified Solomon. At this point we knew everything would be ok and left the newlyweds (all three) to their evening acquainting. Dad told me the next day that he drove out to check on Solomon before sundown, just in case he needed any help, but that the three lions had settled down in the grass. Solomon had never been on grass before either. Often Solomon would roll and rub on the green grass. Life was full of new adventures for Solomon that year.

The zookeepers from Solomon's previous zoo came to visit Solomon and remarked on how happy he looked. Solomon had put on weight, filled out nicely, and his transport bruises had healed up. Solomon was positively strutting as king of the pride. The zookeepers informed us that they thought Solomon was 18 years old. That is really getting up there for a cat. Another visitor from the Humane Society told me that he had rescued Solomon from a pet shop in 1972 and sent him to the previous zoo. That would make Solomon ancient as far as big cats go. No matter how gray in the muzzle he is, Solomon is fat, happy and has a virile twinkle in his eye for an old man.

Chapter 44
Gramma

Gramma's life was an interesting one. Born Elizabeth Alice Cosh, Gramma was one of ten natural children of her parents, Thomas and Anne Cosh. The Coshes had emigrated from England in the late 1800s. If you asked Gramma how many siblings she had, she would tell you she was one of twelve children. One sibling was a stepsister, and another child was adopted into the family. The Coshes were a robust and close farming family. Gramma often told stories of her early days of farming with horse drawn equipment in long skirts.

My grandfather, Ralph Space, traveled the local area trapping for the farmers that needed his services. He was, by all accounts, a character, maybe even a scallywag. Family rumors have Ralph as a moonshiner, outrunning the revenuers in a specially rigged tanker fire truck. Ralph did not like the fact that electric streetlights were installed in the local town of Branchville (6 miles away). After all, electricity is unnecessary and emits unnatural light. Ralph rode his motorcycle through the town and shot out the streetlights with a pistol. Let's just say his reputation as a young man was not the best. Later he would be known as a visionary entrepreneur. It is amazing what time and maturity can do.

On one of his trapping trips to the other side of the town of Sussex, Ralph met Lizzie (as family and friends called her). Lizzie Cosh was the youngest daughter of farmer Thomas and wife, Ann Cosh. It's easy to see how a charismatic, charming, swashbuckling young man could woo a quiet, unimposing daughter of a farmer. They fell in love and were married in the spring of 1925, living with Lizzie's family. In December of 1925, Ralph, Lizzie Space and their one-month-old daughter, Loretta, moved back to Beemerville at the base of Sunrise Mountain. Ralph worked in his garage under the apartment they rented.

Beemerville at that time was a thriving town, with a school, country store, church, and a hotel. A year later a second daughter, Edna, was born. In 1927 Ralph and Lizzie Space bought a bungalow on 1/4 acre in downtown

Beemerville. Fred Thomas Space, my father, was born in a chamber pot on September 28,1928. Tiny Fred was a mere three pounds at birth and my Gramma often told of holding him in the palm of one hand. Loretta, then 3 years old, assisted Fred's birth, as Ralph was out trapping (again) to augment the family income.

Gramma told us stories of early Beemerville, the ladies in long hoopskirts and corsets, horse drawn wagons, tractors, card games, the good old days before refrigeration and electric lights. One night Ralph brought in a poached deer for Lizzie to put up (can) for the family to eat. Lizzie was busy canning the venison meat, when two state police came to the front door and asked for Ralph. Lizzie was in a tizzy. The poached deer carcass was in pieces on the kitchen table. She quickly told the troopers to wait while she checked the animal yard to see if Ralph was back from his trapping, when she knew he was not. Not sure what to do, Young Lizzie took the entire deer carcass to the bedroom and shoved it between the mattresses, made the bed up and perched my infant father on top of it. Returning to the front door, Lizzie informed the troopers that Ralph was not back yet and in true English hospitality, invited the troopers in for coffee. Just as the three were settling down with their coffee cups and cookies, Ralph burst through the door. Momentarily stunned, Ralph looked at the peaceful domestic scene, puzzled. "Where's that deer?" he exclaimed, "I promised these troopers the hindquarter!" Nervous laughter erupted as Lizzie told Ralph and the troopers what she had done. The troopers left with their hind quarter. I've heard this story many times, every time with a twinkle in my Gramma's eye.

Ralph held an assortment of jobs. One job was to drive the silk truck from Newton to Paterson, NJ. Gramma said at this time Ralph started to wear jodhpurs and holstered guns on his hips. Silk was big business in the 1930s; Newton had a major silk factory that shipped silk to the processing plants in Paterson. The shipment trucks were robbed on a regular basis. Ralph took the dangerous job, against Gramma's wishes.

Gramma was by no means a shrinking violet. To assist her husband's growing animal farm behind the garage, Lizzie would can horsemeat (home canning, you know, in mason jars) for the animals to eat. Animal infants (from the animal yard outback) that needed human intervention were brought in for Lizzie to raise. The local farmers and neighbors brought orphaned or injured wild animals for Lizzie to raise. Bobcats, otters, mink and foxes were commonly found in the Space home.

Soon Lizzie had her hands quite full. Three children under the age of five, animal infants warming by the stove, canning, cooking and keeping an eye on the garage while Ralph was out filled her days. And what was Ralph doing? Trapping, hunting, excavating Native American graves and philandering.

Lizzie decided it was easier to set up a kitchen in the garage and gradually her life revolved around the garage kitchen. The original Beemerville Country Store went out of business. Lizzie started to stock the necessities in the garage. Soon Lizzie's store became the center of Beemervillian life. Coffee was always hot and strong, home baked cookies or pies ready for purchase.

Ralph and Lizzie separated. Ralph moved into an apartment over the slaughterhouse. In 1935, this was a shameful situation. Elizabeth Cosh Space held her head high and walked tall for the rest of her life, never divorcing the man she loved, the father of her children.

The small business thrived and Gramma would hire local women to help in what was now a restaurant, country store, Native American artifact display (collecting was a hobby of Ralph's) and zoo entrance rolled into one building. Ralph was in charge of running the fox farm, and the collection of various animals in the back of the building. Lizzie ran the store, kitchen, and pumped the gas out front. The fix-it part of the garage passed by the wayside as ranch raised fur farming became more profitable for the Space family. Then the depression hit.

Local people could not always afford the items they needed from Lizzie's store. Gramma was warm-hearted and did not want to see her neighbors suffer, so the Spaces started to run tabs for the locals. When people could afford it they would settle up their bills. Once a month Lizzie would load up the kids into the Model T and ride around the countryside collecting bills owed to the Spaces. If the locals could not pay, they brought in family heirlooms as collateral or trades. These items were hung on the walls, waiting patiently for their owners to pay up. Sometimes they did not. Ralph and Lizzie had the beginnings of a museum. We also have World War II ration tickets in the current museum from this time. When times were really tough or Ralph wanted to buy another contiguous farm that became available, Ralph and Lizzie would borrow money from Thomas Cosh, Lizzie's father, or other members of Ralph's family. This was an embarrassment for Lizzie, who would in future years be very helpful to her family in an effort to make up.

Meanwhile, the farm and zoo grew with the help of Ralph and Lizzie's three children. Farm families worked, everyone worked. The kids grew up, everyone participating to the best of their ability. Farmwomen grow up multitasking, at least that is what we would call it today. If a job needed to be done, the job was done, by whomever was available. Farmwomen are the original liberated women, at least on our farm. Driving tractors, trucks or backhoes, baling hay, mending fences, shovel work, are all jobs that any woman could do, if she is taught how. Gramma taught my aunts, Loretta and Edna, to drive the "critter" truck. The girls would pick up dead cow or horse carcasses from local farmers to bring back to the farm. Ralph or Fred would skin them out, processing the meat for the animals. These are the type of "we can do it" women my Gramma Lizzie raised.

Gramma Lizzie continued to work in the country store. In 1966 when the small country store became too small to handle the number of visitors the zoo was attracting, a new building was planned. Gramma was integral in the planning of the new kitchen for the restaurant. When progress slowed on the construction of the new building and it appeared it would not be done in time for the school children visitors in the spring, Gramma made her own plans. Gramma ran thousands of school kids right through her dining room and out the back kitchen door of her private home to the zoo proper. On the way back to the school busses, the kids would stroll through Gramma's living room to purchase souvenirs. This happened for three whole months Monday through Friday.

Ralph started the Beemerville Fire Department. Gramma was a founding member of the Grange of Wantage and Ladies Auxiliary, all in her spare time. She worked every day, often from sunup to sundown. She took four or five vacations in her entire working life. Working was her life, always surrounded by family and friends.

After the new building was completed, Gramma continued to rule the kitchen with a velvet fist. The new building had huge glass windows. November first every year Gramma would hand out free ice cream to every kid in town, if the front windows did not get soaped on Halloween. Gramma was a stern taskmaster; any job needed to be done correctly (her way). Somehow, through it all, everyone loved her. Everyone who had ever worked with her called her Gramma. Every teenager that went off to college, after working at the restaurant, would inevitably come home to visit Gramma. She retired from the restaurant in 1992. Every morning the family would start their working day with a breakfast at Gramma's. Gramma would still come down to the restaurant for lunch, feed Snowflake the owl and visit with family and zoo visitors. She would sit at the zoo admission desk, often dosing off in the late afternoon.

THE ZOOKEEPER'S DAUGHTER

In the spring of 1997, Gramma and Snowflake were not doing well. Snowflake died, three months later Gramma took to her bed, spent three weeks of ill health, and died in her own bed on August 1st, surrounded by the family she loved. The entire community grieved. The funeral was packed with people, standing room only, a tribute to the extent of her love and esteem from the community. Floral arrangements overflowed the room. At the head of Gramma's coffin, Snowflake sat perched in all his stuffed glory, a vigil tribute from the animals in Gramma's life. Gramma's worn out body was carried by horse and buggy through the town, around the farm and laid to rest in the cemetery, next to her husband, across the street from her home. Her spirit lives on in all of the lives she touched. I miss her.

Gramma Lizzie and I check out Blizzard,
the African Atlas lion cub.

Gramma Elizabeth ran the kitchen at Space Farms.

Chapter 45
This Little (Piggy?)

Space Farms Zoo has a herd of Collared Peccary. And what, you ask, is a Collared Peccary? The collared peccary is not a pig. 70 million years ago the Suidae (pig family) split, the pigs developing in the Eastern Hemisphere and the peccaries in the west. The peccary's common names—musk hog (due to their musky smell) and javelina (due to their razor sharp tusks)—describe the animal well. The collared peccary is a herbivore found in the southwestern United States and Mexico. They have tough tummies and can digest cacti, even the prickly parts. The collared peccary is a small, 30 to 60 pound hoofed mammal with dark brown /black hair with silver tipping. The collared peccary gets its name from the white collar around its neck. The peccary appears to prance, while they walk or run. Certain foot bones have fused together over the eons and the peccary actually walks on tippy toes. Fast and agile, this wild boar can out-maneuver most of its enemies. Peccarys' short straight tusks fit so tightly together that they actually scrape each other as the mouth closes. This is Nature's way of keeping the tusks sharp. Peccaries are a herd animal, each tribe consisting of 5 to 15 individuals of all ages. The herd communicates with grunts, purrs, squeals and chomping of the tusks. After a 144-day gestation the babies are born in the spring. The female does not make a nest; she simply stops for a moment and gives birth standing up. The babies are miniature spotted copies of the parents, on the move and running within minutes of birth. As much as the parents are just plain old ugly, the babies are adorable. Perhaps it is the young ones' insistent personalities that have won me over. The young prance around the mother with pesky mosquito-like movements, constantly begging for a quick drink. To watch a herd of collared peccary is like watching a three-ring circus, with babies running all over and in between the legs of the adults. Miraculously the young are never stepped on. It amazes me. If the herd is threatened the babies run to the center of the herd, the adults protecting the youngsters.

Late September of that year, the Space Farms collard peccary herd had a set of twins born. September in New Jersey can be chilly at night. The mother lost one twin a few days after birth. Due to the cold snap, Dad asked me if I was willing to raise the other baby. The guys caught the baby and brought the smaller-than-a-football-sized peccary in to me. I set her up in a bed of ground corn and hay. While feeding the "pint sized" peccary I noticed what I thought was a slightly healed puncture wound. It looked like a belly button on her back, only it was oozing a foul smelling discharge. I brought this to Dad's attention and asked if he thought I should take her to the vet.

"You can, but you'd be embarrassed," Dad laughed and said. Dad explained to me that the "wound" was actually a scent gland that the peccary is noted for (musk hog). *Good thing I asked first.*

The little sow bonded quickly to me and was soon following me around. I called her This Little. Since she was not technically a pig, I left the rest of the nursery rhyme off. I found the little sow to be a problem as I tripped over her constantly. She was doing what came naturally, as she would have been following her birth mother. After a few days of tripping over This Little, I put her in the nursery. This Little did not like that at all and would squeal like crazy whenever she heard my voice. She was at my side often.

This Little was not exactly cuddly. Her bristly hair, sharp hooves, nasty tusks, constant rooting for milk and dead duck smell, would not endear her to the average person. Her adoration and devotion to me (now come on, I know it was the bottle in my hand) won me over. This Little stayed with us for two weeks, then suddenly without known reason, died during the night. She may have been ugly and foul smelling to us humans, but I know she will greet me with her piercing squeal when it is my turn at the Pearly Gates.

Chapter 46
Inca the Macaw

A young couple purchased a baby macaw bird around Labor Day in the fall of 1997. This young couple had no idea the amount of love and work a young macaw entails. The couple had seen the newspaper articles on the Atlas lion cubs, born the previous spring, at Space Farms Zoo and my involvement with rearing the cubs. The couple decided they could not handle the responsibility of rearing an infant macaw and gave me a call, shortly after Columbus Day, to ask if the zoo (i.e. me in the nursery) would be interested in taking in their Military Macaw. Dad and I conferred, discussing caging and care. We decided I could accept the bird, if it was healthy, and I would rear it. I would keep the baby bird in my house till the next spring when the weather improved and an outside enclosure could be built.

Macaws are native to South America, Central America and sections of Mexico. They range in size from twelve inches long to three and a half feet long. The colorations are remarkable, bright greens: reds, blue, yellows and even purple-colored feathers adorn the birds. There are many types and hybrids of macaws. The little bird the couple brought in to me was a Military Macaw, with brilliant shimmering green, fluorescent yellow and blood red coloration.

Macaws do make excellent pets if you have the time, space, and patience to deal with the peculiarities of the bird. The beak is made of hard keratin (the stuff fingernails are made of) and is capable of breaking a broomstick. Watch a Macaw eat a Brazil nut and you will realize the strength of the beak. The beak is used for eating, climbing and defense. I can attest to the severity of a macaw bite. If a macaw does not want to do something, it will let you know that, in no uncertain terms. Macaws love to chew on anything—my house bears the scars. The feet are covered in gray, white, or black scales. Each foot has four toes, two pointing forward and two backward that assist with climbing. I'm not sure why the claws are called such; I've seen lesser claws called talons. The claws are strong, pointed and used also for defense. Mature birds can fly. I advise all bird owners to use a veterinarian to help trim feathers, beaks and claws. If you take on a bird as a pet you need to address all the requirements of that bird ahead of time. If handled properly, macaws do make great pets; however, I do not recommend this type of bird for people who have small children with inquisitive fingers. Unless handled by a multitude of knowledgeable trainers most macaws become a one-person pet. Macaws will live for upward of fifty years (some undocumented claims say one hundred years) and this needs to be taken into consideration before you purchase a pet macaw.

When the couple brought the bird to me I inspected the baby for clean clear eyes, good shiny feathering, clean narries (nostrils) and vent (rectum). More importantly the baby bird's leg had a band on it, letting me know this was a legal bird hatched in the United States, not captured from the wild. The baby macaw was loving and handleable. How could I say no?

Jackie and I decided to call him Inca, naming him after the natives of the Central America region. It was a unisex name, we did not know what sex the bird was (and we still do not). The sex of the bird was not important to us; if we needed to know we would have to take him to the vet, as macaws are not sexually dimorphic (exhibiting exterior sexual characteristics). I didn't care what sex it was, I was immediately in love. Inca settled into our home and routine, picking up words commonly spoken in our house. The first word was "Hello," the second word sounds like "Ma." Inca has picked up many words and phrases since then. In the daytime Inca was kept, for his own safety, in a cage. Nightly, after supper, Inca would come out of his cage behind Doug's easy chair and have the freedom to roam the house. Inca always wanted attention, seeking out Jackie or myself for cuddling or petting. Often Inca would sit on our laps as we watched TV, rolling over on his back for additional scratches and petting.

Inca now lives at the zoo in the large flight enclosure Dad and Doug built for him, on the hill. I play with him daily. Inca can see me as soon as I walk out the kitchen or zoo entrance door and always greets me with, "Hello Ma," asking for "Scratches." He practices his speech and squawks constantly. Inca is a cherished addition to my life.

Chapter 47
Oh Deer, a Broken Leg.

Many things had changed at the zoo while I was living in Pittsburgh, including what we fed the carnivores at the zoo. While the family ran a mink ranch, the animals at the zoo were fed chunk raw meat or poultry products before they were ground up for mink food. In 1987 Space Farms went out of the mink farming business. Dad developed an alternative food source for the carnivores and omnivores. The state of New Jersey has an overpopulation of whitetail deer. The NJ wild deer is in close proximity to the human population and human activities have made them less afraid of cars. Numerous deer are hit by cars every year. Dad developed an information network of state troopers, school bus drivers, county workers and business commuters. Everyone knew to give Fred Space a call if they saw a road killed deer in the morning. This system worked out well for everyone. The yucky road kill was picked up, keeping the community clean, preventing the obvious health hazards and wild predators from coming too close to suburbanite homes. The fresh useable carcasses were recycled as zoo animal food. This gave the animals at the zoo a completely natural food source, no artificial coloring, preservatives, smell retardants etc. This was no small daily undertaking. In the year 2000 Space Farms picked up 2000 dead deer off the roads.

The road kill pick up job was really yucky. No other words can describe it better than Reader's Digest's worst jobs list. Road Kill pick up was in the top 5. (Do you really need to ask why?) Dad does that job daily, even though he claimed to have retired at age 65. My brother Parker has often remarked, "Other fathers retire and go golfing or fishing, our dad picks up road kill." If Dad went on vacation or observed his religious High Holidays, Doug, Parker, Bill or I would fill in.

Our family observed the traditional Christmas and all the other normal holidays of the year, though we were usually working (animals need to eat every day, even if the zoo was closed to the public). But, without question, you ask the men of the family which is the best holiday of the year and they will inevitably reply, "The First Day of Deer Season." Always the second week of December, this religious High Holiday was anticipated with excitement. Weeks of spotting deer, planning the drives, preparing the rifles, dusting off the glowing orange hunting clothes, planning to take the day off of work, cleaning the meat smokers and freezers and talking of the weather (a light dusting of snow is preferred for tracking) filled the thoughts and minds of the hunters. The womenfolk cleaned out the freezers in preparation of the new meat coming in to feed their families for the year.

In years past Gramma would be up before dawn to prepare a huge breakfast for the hunters. Not only hunters of our family but a number of family friends would stay in our homes to participate in the big day. Breakfast would be served by 5am in order for the hunters to be in position in the woods by dawn. With Gramma's death, there was a gap in tradition. I had hunted as a child, but gave it up when Dad yelled at me for blowing my nose while on a stand. "You'll scare all the deer away!" he had barked at me. Doug is a fisherman, not a hunter. As I've stated before, I am not the best morning person, so I did not volunteer to cook the breakfast. I did, however, volunteer to cook the lunch for approximately twenty-five men. I felt it was my duty to carry on Gramma's tradition. It is a fun day; the festive atmosphere and hunting stories that evolved over the years made the First Day a holiday equal to any other.

The First Day of 1997 I cooked lunch for the hunters in the main building's kitchen. Spaghetti with venison meatballs (got to use up that old meat), salad and dessert served with lots of hot coffee and cocoa was the menu for the day. I had invited two girlfriends to help with the meal. Doug and Bill joined us for lunch. The hunters came in at noon, ate, warmed up, and were gone again by one o'clock. Doug and Bill went back to the daily grind—Bill to fix a tractor and Doug to gut and skin the day's road kill that he had picked up that morning. After the men ate, the girls and I sat down to eat. Usually I ate with the men, but the increased number of hunters for lunch kept me busy throughout the hour.

The ladies and I were finishing up our lunches, having a gabfest, when one of our neighbors popped her head through the front door. I had seen her check her mailbox near the road through the large glass windows of the main building. "Doug said he wants to see you," she said nonchalantly.

"OK," I replied, thinking nothing of it. I continued talking to my friends for a few minutes. In the back of my mind a small thought made its way to the surface. *Hmm... Wonder what Doug wants. The last time I was called to*

THE ZOOKEEPER'S DAUGHTER

the slaughterhouse Dad showed me a doe with quadruplets in her womb. Time before that was another oddity, a deer with two hooves on one leg. Wait a minute, I pondered, *this is December, fetuses would not be that developed. Hmm... better go check.* I told the girls my thoughts and left to cross the street to the slaughterhouse to see what Doug wanted. He was working alone; maybe he just needed two minutes of help.

As I walked across the street, I saw Doug half-leaning, half hanging onto the back of the death mobile (the truck we used to pick up the dead deer) parked in front of the slaughterhouse. Something was wrong, Doug looked awfully pale. *This must be a gem,* I thought, *Nothing turns Doug's stomach.* Doug lifted his head and hollered to me, "Get your car, I've broken my ankle!" *OH S**t!* My heart screamed, as I ran back across the street to get my car. I dashed into the building, handing my friends the keys to the front door of the main building. "Put what you can away in the cooler, and lock up," I commanded. "I've got to take Doug to the hospital, he thinks he broke his ankle." The girls looked at me and accepted the keys.

I jumped in my Jimmy and drove the twenty-five yards to the slaughter house steps, backed the Jimmy up the ramp and helped Doug get into the car. No use calling 911, all the ambulance drivers were out hunting with my dad. By the time they came in from the field I could have Doug at the hospital. As I drove the six miles to the hospital, Doug said he felt like he was going to pass out. He lowered his seat to the resting position but temporarily stayed lucid. I did not purposefully hit all the bumps that Doug was yelling at me for. I knew he was in excruciating pain. Doug passed out two miles from the hospital. I nervously ran a red light, after seeing no other traffic. I prayed for a policeman, but you know what they say...

I pulled into the emergency entrance of the hospital, went inside to get help and a wheel chair. The nurses and orderlies came out to the car and looked at Doug. "Oh MY," calmly stated one of the nurses. And I looked to see what she saw. Doug's coat, jeans and boots were covered in blood, and pieces of intestinal goo. The nurse, knowing that it was the First Day, asked, "Where was he shot?" I must admit I laughed at that point, Doug was lucid again, but not laughing at all. I explained to the nurse that the blood and intestinal goo was not Doug's but belonged to a dead deer. Doug had fallen, tripping over a box, while dragging a dead deer backward into the slaughter house, the deer fell on him, and he in turn, on top of a box of (what else?) dead deer guts. The nurse and orderly looked at each other, wrinkling their noses as the reality of the smell confirmed my statement.

"OK," said the nurse, "let's get him out of the car."

"Before he passes out again," I added. They looked at me and got to work.

Doug had done a real job on his leg and ankle. When he tripped over the box of dead deer guts, he pirouetted, twisting his ankle, then jamming the tibia and fibula on different sides of his ankle to the floor. Then he fell, dragging the deer down on top of his damaged leg. The Doctor said he had only treated one other case like this and that was in a professional football game. Doug was in pain, big time.

The nurse offered Doug a shot of painkiller as they attempted to cut off his pants. Doug is very reserved and refused to let the emergency room personnel cut off his filthy jeans. Doug was, at this point, in shock and very uncooperative. He did agree to let them cut the laces, but not his new boots, to get his boot off. As they pulled off the boot, Doug went through the ceiling. Again he was offered a painkiller shot, but refused to take off his pants in front of another woman, other than myself. *Isn't that sweet, very dumb, but sweet.* Doug is usually the salt of the earth as far as nice is concerned, but he was having a really tough day. I first suggested that the nurse shoot him through his pants. The nurse looked at his pants, then looked at me.

"With all that goo? He would get an infection for sure!" Doug spoke up and suggested that we cut a hole in the thigh of his pants and inject him there, because he was, "not going to take off my pants!" That worked. As soon as the medication took effect, Doug was punch drunk. He was a riot, making raunchy jokes, pinching the nurses. I apologized to the nurses, but we all agreed that this Doug was more handleable than the mean, nasty, hurting man who was in the bed a few moments before. The Doctor and nurses had the situation under control. Doug was shipped off to x-ray.

I stepped outside to call my Aunt Loretta, who lived up the street from the farm. I asked her to get my daughter off the bus, and go inform Dad, who was still unaware and hunting a field. Aunt Loretta accepted the emergency orders, and took care of Jackie till I got home later. Doug came back from x-ray, the news was worse than we had thought. Doug had broken two bones at his ankle, dislocated his lower leg bones, jammed them to the floor over his anklebone, and broken his tibia below his knee. He would need an operation. To make matters worse, he had just eaten and would not be able to be operated on for eight hours. Doug was doing fine, now sitting carefree in his underwear, ogling the nurses, sailing high on the painkilling drugs.

The Doctor decided it was time to set Doug's leg. I started to leave the room, figuring I was only in the way. The Doctor stopped me. He explained to me that I was needed to put my hand on Doug's chest. I knew I could not hold him down if Doug decided to come off the table and told the Doctor he was better off with another orderly. The Doctor said, "NO, it's your job, a husband will behave more macho and suck it up more for the wife than for an orderly. This is going to hurt." Machismo is multi specie, I should have known. I tentatively put my hand on Doug's chest. My reserved husband had been stripping, with sound effects, for the nurses after the third shot of painkillers. By the eighth shot of painkillers, Doug was spread eagled with orderlies and nurses on each appendage. The Doctor and an orderly pulled his broken leg out straight, giving the ankle a quick jerk. Doug arched his back, his eyes rolled to the back of his head, but like the Doctor had predicted, did not come off the table, as the room filled with the sound of popcorn. The leg was set, and wrapped. All to do now was to wait for Doug's spaghetti lunch to digest.

I drove home to let Jackie know what had happened. I thanked Aunt Loretta for keeping Jackie. Next we stopped and filled Dad and Mira in on what had happened. We ate supper with Dad, then headed back to the hospital for the eight o'clock operation.

The operation went fine, many thanks to a great doctor. Doug was in the recovery room, coming out of the anesthesia, still cracking bad jokes. He was the first stand up comic I ever saw laying down. My hubby came home on crutches two days later. It was a tough couple of days, but soon life settled down. Doug was settled into his easy chair, Jackie off to school every day before I went to work, at eight a.m.

Dad was still in hunting mode. He deserved every minute of hunting joy, as he had worked hard all his life. Bill Raab had had an ankle replacement the month before. Parker had the rest of the farm and zoo to run. I received a few minutes of instructions on the winch of the death mobile, was assigned the dead deer pickup routes, and some of Doug's jobs, on top of my own. Talk about a challenge, my life was full. The dead deer pick up job was in itself, yucky. Sorry I can't think of another word. But there was a certain joy in cruising the highways and local roads, blaring the radio, singing at the top of my voice. I had special gloves and boots to protect myself from ticks, lice and you know... germs. The best part of the day was when I pulled into the slaughter house with my load of up to eighteen does and some really big bucks (my best day). I met Dad's hunting party one day as I pulled in and we compared bucks... I had bagged the biggest one! The job seems gross and is in many ways, but I rationalized that I was grocery shopping for the zoo. As Christmas was approaching, I will admit that on company time I drove into the parking lots of retail stores with my dead cargo in the back of the death mobile. You can imagine the looks I received from other shoppers as I walked into the stores. My days and nights were very busy.

Inca the Macaw was still at home, a four-month-old infant. His cage was situated behind Doug's easy chair on the only spot of stonework in the living room. Inca was a loving baby. I would take him out nightly, sitting with Inca on my lap, petting, and stroking the bird. I enjoyed the bird after a diverse day of work, which now included picking up dead deer. Inca would squawk, coo and cuddle, relieving some of my tension. One day as I came in from work, I heard a lot of squawking and Doug's cursing. I hurried to the living room to see Inca perched on top of Doug's chair, nipping at Doug's bald spot. Doug was livid. The bird was upset. Inca had escaped his cage, walked to Doug's crutches, climbed up the crutches leaning on the side of Doug's chair, knocking them to the floor, finally perching himself above Doug's head on the recliner chair. Inca wanted loves, and Doug was not in a loving mood after three hours of bird shenanigans. Doug was trapped in his chair, could not get to his crutches, and was at the bird's mercy. And Inca wanted more attention. I put better clips on the birdcage to make sure that did not happen again. Doug still hates "that Macaw."

On the nineteenth of December I was in the office filling out the dead deer reports, when Dad caught me with a surprise. Two little lion cubs had been born. Dad said, "Don't worry about them, we'll let nature take its course, you've got your hands full." I looked at Dad, gave him a wink and said, "That's the best news I've had in a month. Bring them in if you think they need me." The temperature had dipped. It was, after all, December. Dad was worried about the babies. The mom was not keeping them warm, so he and Parker brought them in to me. I named them Holly and Ivy. I enjoyed the distraction, the cuddling and feedings. OK, I just love little lions. I put them on scheduled bottles and hoped for the best. They were good-sized being the only two in the litter. They did well. Christmas came and went, Doug still with his leg up in his chair, Inca trapped behind it (until I came home and let him loose in the kitchen), Holly and Ivy growing fat in the dining room in cardboard boxes, and Jackie helping me cope with it all.

After Jackie had gone back to school, I was on dead deer runs again. I would pace myself to be home for cub feedings and check on Doug. It was still hunting season. I was delayed one day by a bunch of hunters that pulled up to the dead deer I had on the winch at the back of the death mobile. Seems they had shot the deer, it did have a bullet hole, but I thought it had been left abandoned. I was running late on my 10:30 a.m. feeding for the cubs. The deer was just about in the truck when the hunters decided to ask for the deer. I snapped, "First come, first served!" One look at my face and the hunters decided to go get another deer. You don't mess with an adoptive mother lion.

I got home late, 11:30 a.m. with visions of screaming lion cubs, and Inca pestering Doug as I walked through my back door. I had seen two familiar cars in the driveway, I knew who was inside. I was amazed as wheelchair bound Jim (with spinal bifida), Bill (with crutches due to his ankle replacement) and Doug (with his broken leg) greeted me. "What's for lunch?" they asked incredulously. Overwhelmed, I was still tense from the hunters and the dead deer episode and being late to feed the lion cubs. I broke down with laughter at the sight of the three cripples, two helpless lion cubs and a macaw hollering, "Ma Ma" in my living room. I had spent the morning scraping dead deer off the road and the men wanted what from me? I don't remember exactly what I said but I'm sure it was a gem! I fed the lions, the men, then thankfully had work to go back to. Work can be cathartic.

Holly and Ivy grew fast. Holly went to a couple in Colorado. Ivy stayed in the house till spring, using a plastic sweater box for a litter pan. Winter is an off season; no one wanted her... yet. I enjoyed having her around the house, even though the smell of lion urine permeated the house. You can imagine if you've had one cat in the house, multiply the smell times five, that's how big she was by the time spring came around. Ivy was full of life and romped through the house, a twenty-five-pound, playful kitten with the agility of a bull in a china shop. Our house cat, Daphne, quaked in apprehension at the sight of her and lived most of her days upstairs if Ivy was roaming free downstairs. I was so happy to see spring come the next year. Doug went back to work in April, after the doctor released him from his medical leave. Life seemed to get back to its uneven keel.

Chapter 48
Santa May be Late This Year.

Santa may be late this year,
at the rate we're picking up dead deer!
Road kill is the zoo's main diet
Oh, don't knock it, I've even tried it!
Yes my friends you won't believe
this latest new talent I've pulled out of sleeve.
I am now the pick up Queen,
just give us a call, I'm on the scene.
An how you may ask did I get this duty,
to pick up the roadside's bountiful bootie?
Dad's gone a hunting (yes again)
and Doug, remember him - my hubby and friend,
took a trip backward, and then went down
under a dead deer on the ground
broke his leg, ankle and worse
dislocated it all, boy did he curse!
And on the First Day of deer season, what nerve!
So I've been asked to step up and serve.
It is a long standing family tradition,
If you want the job done, give it to the women.
Such talent, amazing to be so diverse
and then to sit down and put it in verse.
I've been a libber, but NOW am not so sure
if I want to do *that* anymore!
This past year has flown
so many babies have grown,
six lions, eight fawns, a macaw and an owl,

THE ZOOKEEPER'S DAUGHTER

seven turkey chicks (I still hate that fowl).
And Jimmy the cutest little llama,
who because of a boo boo wore pajamas.
I gave speeches and talks, did the PR walk,
was on radio and TV (that fat wrinkled lady cannot be me!)
My babies are placed all over the states
to visit on the next trip I make.
Jackie is growing - my sweet, smart girl
starting to take on the whole world.
One day a tomboy climbing trees,
The next all dolled up pretty as you please
For Christmas she got a feather boa
Now she "does" Marilyn you know.
A wonderful child, every mom's dream
one year and then the terrible teens!
Doug - he's recouping, leg in the air
at home nice and cozy in his easy chair.
My dad is fine, hunting every day
and on August 1st my Gram passed away.
She was taken by horse and carriage once around the town
and put to rest on family ground.
Everyone turned out, a sight to see,
a fitting tribute to a grand lady.
She smiles over me as I make my way
working the farm each and every day.
This year was good, attendance was great,
we live all year on the summer's gate.
'Twas six days before Christmas and, oh, surprise!
two little girl lions materialize.
Holly and Ivy, they sure are sweet
as they pitter and patter on their tiny feet!
That's my life, silly but true,
with all the variety of things I do.
Not to worry, though, Santa's deer are fine
all those I pick up off the yellow road lines
are the ones that never learned the rules
And flunked out of reindeer flying schools!
Hope Santa is good to you. Love, Lori

Chapter 49
Pebbles, Brooke and Rocky, the Fishers

The fisher is a woodland creature that few people have seen in the wild. Their shy, elusive and quick nature makes them difficult to spot. The fisher's dark brown to black dense fur easily blends in the forest under brush and swamplands, where it prefers to live. A cousin to the weasel and mink, the fisher's body is long and cylindrical. The male weighs approximately 20 pounds, while the female is much smaller at 4 to 6 pounds. The fur varies in color from tan to jet-black with the majority of fishers colored a flowing chocolate brown. The fur around the face is lighter in color.

While the fisher used to roam the entire northern part of the North American continent, they were trapped out of the northeastern states by 1920. Trappers considered them an elusive prize, the pelts bringing $345.00 a piece. The price of two fisher pelts could buy you one of those new fangled cars in 1920. With proper conservation efforts by federal and state governments and the cooperation of trappers, who helped to live trap and restock areas, fishers have been spotted on the New York/ New Jersey state line. This is an unheralded success story, as so few people know about fishers, and fewer people care. It is one of the reasons we kept fishers at the zoo. They are not much to look at, large fuzzy brown critters, sharp teeth, beady brown eyes, speedy, but they do have cute teddy bear ears. Fishers live within a home territory of 10 square miles, the size of the territory varies with food availability. Contrary to their names, the fisher rarely eats fish. The main part of their diet consists of rodents, beaver, muskrats, woodchucks, birds, eggs, frogs, salamanders, and some vegetable matter.

In the spring of 1996, our female fisher gave birth to three kits. The zoo had fishers prior to our mink ranching days. Dad convinced me that the fishers would make great "pets," as he knew a number of trapper friends who had raised them. It would also be great to take them out to lectures as so few people knew about them. It would also benefit the zoo, as "people friendly" fishers would not be afraid of the public. The adult fishers at the zoo were not hand raised and spent their days hiding in the logs inside their enclosure, coming out at dusk to play and eat. I was game for the challenge. I had no idea what I had gotten myself into...

Pebbles, Brooke, and Rocky were removed from their mother and brought to me at five weeks old. I put them on a formula of Esbliac and fed them with a bottle. It took them a couple of days to get used to a rubber nipple, but they caught on. The girls were so much smaller than the boy was. I probably coddled them a bit more (I'm a sucker for the underdogs). A trip to the vet for general check up and shots assured me they were doing fine. I hand ground venison with dry cat food to add to their diet.

Little did I know at the time that I accepted the challenge of hand raising three little fishers, that fishers can jump up to 5 feet and run like the dickens. Having three young fishers in the house was like having three lightning bolts on amphetamines! Talk about pandemonium. Pebbles, Brooke and Rocky would rest all day at the zoo in the nursery, and by the time they got home at 6 p.m. were ready to rock and roll. Running, jumping, nipping and chattering their way through the house, my home was not at ease. The fisher is an excellent climber and can climb trees forward and backward, head first and tail first. Our circular stairway was a favorite climbing area, no matter if some human feet were also there at the same time. Curtains, sofa, lounge chairs, and our legs all bore the marks of the climbing claws made for trees. No wonder the fisher has few enemies in the wild. Only bobcats, lynx and birds of prey can catch the "faster than a speeding bullet" creatures.

And the smell... Did I mention the smell? The fisher, like other members of its family, has anal scent glands. The musky smell permeated our house. It did not help that the young male was also marking territory. When frightened they will emit a discharge, not exactly spray like a skunk, but it was smelly. No, correct that—they stunk. Their musky smell is similar to Swiss cheese left in the sun, an intrusive insult to the sinuses. Keep in mind that I have deodorized skunks, and dealt with some very smelly poop from many different species. Fishers top them all as far as smell goes.

Over the years my family has made many life long friends. One of my grandfather Ralph's and my dad's friends is Leonard Lee Rue III. Mr. Rue is my dad's age and is affectionately called Lennie. It seems so strange to call such a distinguished and accomplished photographer by a boyhood name, but that's how close the family is. Dad had spoken to Lennie, just shooting the breeze as friends do. Dad mentioned that I was raising fishers. Lennie was very excited about photographing the fishers. The fisher is so elusive in the wild that even a famous expert wildlife photographer like Lennie had not had many chances to "shoot" them. Lennie came up and we took the fishers out to the woods for photos. It was good to see Lennie again. I had not seen him since he had taken pictures of Moses the lion cub and myself in 1974. He was as handsome as ever, his smile emanating the warmth of years of friendship between the families. It was fascinating to work with Lennie, watching an expert in a field I had no knowledge of. We exchanged animal tips on the fishers' behavior, in comparison to other species of the same family such as the mink and the weasel.

It is at this point that I must admit my psychological failure to bond with these creatures of God's kingdom. I took these animals on, but could not stand the smell. They were friendly, perky, inquisitive, intelligent, interesting and a great hit on the lecture circuit, but I couldn't get past the fact that they looked like a mink, and smelled musky like a mink. So many years of pelting ranch-raised mink had tainted me to their special "attributes." They were moved out of the nursery and into the zoo proper in the fall. These were the only animals I was ever glad to be rid of. Pebbles, Brooke and Rocky are doing well to this day. They respond to the public and still recognize my voice. I visit them just to remind myself that I am not the perfect Zoomama, I just could not love them as I had so many of my other babies.

Chapter 50
Ivy Takes a Road Trip

Ivy the lion cub was moved to the zoo proper in April after the weather broke and spring had sprung. She was a huge kitty by that time, a whopping thirty-five pounds, about two and a half feet long (not including tail). I was so ready to move her out of my house. The nursery in the zoo had an available enclosure that would do for the time being. Because she was born in the winter (the off season for zoos), she had not been placed yet. She was affectionate to the extreme because she was in the house so long playing with Doug, Jackie and I. I needed to find her a special home. Dad and I had discussed putting Ivy back in with the parents if we could not find a home for her. Trainers want younger cats that can bond with them. Ivy had definitely bonded with my little family. She was doing fine, growing more and more every day.

I received a phone call from the distraught dad, Randy, of Rose and Petunia (two female cubs of the big litter of '97). Petunia had had an accident, cracked her skull and died of complications a few days afterward. Rose was depressed and was "off feed" (a zoo expression for not eating up to normal par). Did I know of any other lions around that he could buy, trade for, or would someone donate? Rose needed a friend. Randy's sanctuary was the type of place I wanted for Ivy. Randy's heart was really into animals, and he had a good supply of food and sponsors to ensure the future. I told Randy he was the answer to my dreams for Ivy. Ivy was six months younger than Rose, but should get along just fine. Randy agreed to keep an eye on Ivy to make sure Rose did not bully Ivy around too much. I had no doubts that Randy would keep his word.

Randy was making a run into New York City with his eighteen wheeler in a couple of days. I agreed to take Ivy to the truck stop off of Route 80, so Randy did not have to drive his big truck all the way to the zoo on the windy country roads.

I made arrangements for Mom to watch Jackie while Doug and I took Ivy to the truck stop. Randy would be there around 9 p.m. After work on May 18th, Doug, Jackie and I came home and ate supper. Mom arrived so Doug and I went to put Ivy in her crate at the zoo. I grabbed some bottled water so Randy could give Ivy water on the fifteen-hour trip home to Michigan. Doug, Ivy and I set off on our hour trip to the Route 80 truck stop. The trip itself was uneventful, though I always worry about car accidents and having to chase a lion across traffic. I am a certified worrywart. Doug, Ivy and I pulled into the truck stop a little early, and waited for Randy. There were a number of eighteen wheeler trucks at the truck stop. We watched the people come and go under the glare of the neon lights. I love to people watch, so you can imagine the fascinating opportunity at a real life truck stop. Some interesting characters paraded in and out of the road weary or sleeping trucks. I wondered what kind of business they were all involved in, not realizing that we were probably the weirdest there, the only ones with a lion cub in the back of the car.

Doug kept asking me how I was, almost to the point of being a pest. We had been married for fifteen years. Doug was usually in tune with my feelings. I thought it was sweet that after all the years we have been together, he was concerned that I would be upset about sending Ivy away. Randy pulled his eighteen wheeler into the truck stop, saw our car, and tooted his horn. Ivy had been watching the lights and the commotion of the truck stop with interest out the windows of the car. She had never seen the traffic or the lights of the "big city." With the tooting of Randy's truck, Ivy snorted and huffed, then meowed to me in apprehension. "It's OK, girl," I must have said for the thousandth time.

"She's fine, you all right?" Doug looked at me and chuckled. I glared at him. *Boy, he is being a pest.* Doug and I got out of the car to greet Randy.

Randy and I signed the paperwork involved on the hood of our car. Randy needed certain papers to cross state lines with the lion cub. I had the paperwork all done except for the signatures and the license plate number of the eighteen wheeler. With some last minute instructions (that I'm sure Randy did not need), and a farewell kiss through the grate of her crate, Ivy was on her way. Doug, Randy and I carried Ivy in her crate to the cab of the truck. Randy would talk to Ivy the entire way home, to start the bonding process exchange.

Doug and I watched Randy and Ivy pull away. Doug handed me some tissues and asked again, "How do you feel?"

"I'm fine," I replied abruptly.

"Are you sure?"

STOP IT, I'm on the verge of tears, don't rub it in! I thought, barely composed. "I'm fine," I said out loud, thinking, *It's nice of him to be so concerned.*

"Just wanted to know." Doug's eyes twinkled. "Just wanted to know how you feel. This is the first time you gave away free (slang word for cat) at a truck stop!"

My husband is such a wise guy.

Chapter 51
Pickles the Opossum

The Opossum is an underrated animal. Most people only see them in the wild as road pizza. Possum are fascinating creatures, first showing up in fossil records during the mid-Cretaceous period, 100 million years ago. The elongated snout, sharp teeth, black beady eyes on a white face, pale pink nose, naked prehensile tail, hairless ears and human shaped feet make them less than attractive, except to another opossum. The fur is exceptionally silky in patterns of gray, silver and white. The fur of the opossum is valuable to trappers. But don't let the looks of an opossum fool you, they are tough creatures. Often losing ears, tail or toes to frostbite, possums do not hibernate, just hole up for a while when the weather is coldest. Possums are very clean, contrary to general belief. An opossum spends a good part of its day preening, just like your housecat.

What makes the opossum special on the North American continent is that the opossum is the only native marsupial. The female opossum has a pouch to rear young. A female possum has a gestation of 13 days, at which time the newborn are about the size of the eraser on a led pencil. Newborn possums are fetal in appearance, eyes sealed shut, hairless, and a tiny mouth, but with gripping front feet. The mom possum does not build a nest, merely stops for a few hours to allow the infants time to crawl from the vagina to the marsupial pouch located on the lower abdomen. Inside the pouch are a series of thirteen tiny protrusions on a flat mammary gland. The minute infants creep to the pouch and attach to the thread-like nipples, they will remain there for two months. Although thirteen nipples are available and often thirteen or more newborns crawl to the pouch, six to eight babies are all a mamma can carry on her back. As with the survival of the toughest, the weak ones are pushed aside. At eight weeks old the possum babies, about the size of a spool of thread, leave the pouch and are often seen traveling on momma's back.

Many stories have possums hanging upside down from trees by their prehensile naked tails. I personally have never seen it. Momma possum travels the forests, fields, and suburbia with her litter clinging to her back. Possums are omnivorous, eating just about anything they come across, including carrion. Nocturnal possums are seldom seen by people, except in the headlights of their cars at the last moment. "Playing possum" is a phrase born of truth. If a possum is threatened it will roll over in a fetal position, play dead with its mouth open and tongue hanging out, often regurgitating its last meal. This is a natural instinct, a nervous paralysis, designed to "gross out" its tormentor, hopefully leaving the possum alone. Neat trick.

Often in the spring, when the young come out of the pouch and are riding along on momma's back, babies fall off and are lost. Momma possums cannot count, and seldom retrieve the fallen young. This is nature's way.

A young possum was brought in to me by a couple that had been hiking. A car had hit the momma. All of the other young were crushed or had wandered off. This little possum was the only one still trying to nurse from his dead momma. I named him Pickles, he was about the shape of a gherkin, three inches long, and like other possums had a sourpuss face. He was so ugly, yet tiny and clean. I set Pickles up in a cardboard shoe box with a hot water bottle, fed him with a baby doll nipple and premature human baby Simalac. He gobbled his five drops of formula every hour and started to grow. Pickles spent his nights in the cardboard box at my house, his days at the admission desk watched over by Alice and Daisy, the ladies who manned the desk. Grandmas themselves, they are great babysitters.

Pickles greeted many busses of school children coming to visit the zoo. Busses usually showed up around ten in the morning, a feeding time. Pickles would hold his own bottle, lying on his back, in my shirt pocket. When he was done drinking, he would crawl out of my pocket, much to the delight of the kids I was explaining the zoo rules to. I really don't think the kids heard one word after Pickles came out of my pocket.

There is no way to explain how cute a baby opossum is. A tiny pink nose, grabbing hands, a gaping mouth that sneered into a smile, and a clean pink tail curled around my finger, gave Pickles an impish gnome appearance. Opossums are so different than our perception of beauty, an antonym with an awe of its own. I know most people thought he was beyond ugly when I took him out to speak to groups at the zoo and on the lecture circuit.

Pickles grew and by the time he was the size of a gerbil he was eating Gerber baby food. His favorite was the fruity kinds. He was, however, given a selection of what Gerber had to offer. He still loved his bottle. I fed him bottles on the lecture circuit all that summer. When Pickles became large enough and was eating venison mixed with dog food, he was moved to the zoo proper to share an open enclosure with a family of raccoons. They got along well, Pickles often napping under a pile of leaves in the sunshine.

    Space Farms has a history of healthy handleable animals and are often rented for TV and movie spots. Pickles had a bit part in a movie. His star never rose higher than that, but he remains a star in my heart.

Chapter 52
Niles and Sahara, the African Atlas Lions

Niles and Sahara were born June 5, 1998. (It took Solomon a little time to catch on to what the ladies needed to teach him.) Two healthy Atlas African lion cubs, and the lioness decided not to raise her cubs, again. We brought the cubs into the nursery. The cubs ate well and progressed according to lion cub development schedule.

I began taking lion cubs out for the zoo public to view the feedings. The demonstrations and discussions on lion cub care were a big hit with the public, often gathering two hundred visitors at a time. As soon as the cubs were past the critical point of ten days old, I took Niles and Sahara out to the picnic table under the pine tree by Gramma's house. No one was allowed to touch the cubs (my germ phobia again).

One day, as I finished my talk on the care, feeding, development and the adult lions, I asked the crowd for questions. From the back of the crowd came a question, "What is that cub crossed with?"

"Nothing, this is a purebred Atlas or black mane, African Lion," I replied. I went on to give all the characteristics of the Atlas strain of lions

"It has to be crossed with something." My inquisitor again piped up.

I scanned the crowd, thinking my dad had placed a heckler in the group. Dad and I play tricks on each other all the time, I like to give kids in his snake lecture audience really tough or embarrassing questions to ask. So this was pay back time, I figured. I looked around for the twinkling eyes of my dad when he is pulling a good joke. Dad was nowhere to be seen. There were no familiar faces in the crowd and I could not see who had asked the question. I continued speaking, explaining the history of the Atlas African Lion.

"If that cub is pure lion, why does it have spots?" The questioning voice was still not satisfied.

"All lion cubs have spots, sir, the spots gradually fade by the time the lion is two years old," I answered. I finally saw who was asking the questions. I did not know him, and he did not have the twinkle in the eye that I would expect from someone playing a practical joke. He was serious.

"Not all lion cubs have spots," he replied. I looked at him incredulously. I had no idea what he was talking about. "Not all cubs have spots, Simba didn't," he clarified.

I could not believe my ears (and neither could the crowd). "Simba, sir, is a cartoon. Maybe that movie company should hire me as a consultant!" I quipped. The crowd laughed, but I should have been nicer to the man.

By the time Niles and Sahara were eight weeks old, I had to take Niles on the dreadful trip to the airport. He was headed for Wisconsin, to a dealer who would place him with another zoo. God knows I hate that trip to the airport.

Sahara stayed home at the zoo and was the star of my lion speeches for the rest of the summer. Sahara's new dad was a trainer, but he didn't want her until she was older. No problem with me, I'd keep all my babies forever, given the choice. We already had three adult lions. It would be wrong to put her back with the pride where she would inevitably breed with her own father. Sahara left me on October 26th. I was happier when I met the new mom in charge of Sahara. This mom could "talk" just like a lion and had lots of experience. The world needed my babies to expand a limited gene pool in captive Atlas African lions. That's just the way it goes.

Chapter 53
The Anti-condom Snake

I love to speak at campgrounds. The atmosphere is casual and always vacationish. Many of the local campgrounds have seasonal campers who I know from the community. I was at Kymer's Family Campground in Branchville, just a couple of miles from home (the farm). As usual I had spoken on a lion, Sahara, an owlet, then brought out the snake. The snake I keep till last so the faint of heart do not leave my speech early. After I finished speaking about the snake, I asked for questions and fielded the answers. A little boy with large, round, innocent eyes was in the first row of the hundred campers gathered. He raised his hand and I called on him. He stood up and in his clear schoolboy voice asked, "Hey, lady, have you ever heard of the ANTI-CONDOM snake?"

I could not believe my ears. I asked him to repeat the question, wondering if this little angel was playing a prank on me.

Still standing he repeated, "Have you ever heard of the ANTI-CONDOM snake?" He was clear and distinct in his pronunciation. The background chatter in the pavilion stopped and a hush fell over the hall. I was at a loss for words, which doesn't happen very often. I studied the sweet little boy and tried to figure out if this was a joke or if he was serious. The silence was awful. One of those moments that you are so flabbergasted you just can't figure out what to say. There were a lot of small children in the audience. I was stuck. My mind was racing for something, anything, to say in front of a mixed audience of all ages. Finally, after a long pregnant pause, an expectant lady in the back of the pavilion rubbed her swollen belly and shouted," I don't know about you, lady, but I have seen it, and it bites!" The crowd burst out laughing.

The young boy teared up, thinking we were laughing at him. Then he got angry and shouted," You know, the ANTI-CONDOM snake, lady, the one they made a movie about!"

Then I knew what he was talking about. "Oh I know," I consoled him.

"The anaconda snake lives in South America. That snake was in a horror movie recently." The young lad smiled, satisfied with his own intelligence and sat down. Funny how the accent on the wrong syllable can throw ya.

Chapter 54
Animal Photography

One of my jobs is to assist professional photographers when they come to the zoo. Photographers from newspapers, magazines, books, and TV shows would come to the zoo to take certain animal pictures. The photographers want to get as close as possible to animals, behind the guard fencing in place for the general public. This was not a problem with non- dangerous animals, although there is a chance of danger whenever you are working with wild, undomesticated animals. Even hoof stock can stampede when scared by a flash or the blinking red light on the front of a video camera.

There are certain spots in the animal enclosures that are exceptionally dangerous. All of the big cats (lions, leopards, cougars, tigers, and jaguars) and the bears (grizzlies, browns, black and Himalayan) had feeding stations. These feeding stations are big enough to put chunks of venison or other large food items in. Any place that a large chunk can go in, a paw with claws can come out to grab you. Animals play rough, even if they intend no harm. We have a number of animals that were not hand raised and will grab you simply because they can. Ask a mountain climber why they climb mountains and they will answer, "Because it is there." Same theory. If you inadvertently lean or rest your hand on the animals fencing, the animal will take that as a challenge to his territory (his enclosure), and come at that hand or body part.

All of our animals are kept with others of their own species, so there are more than one animal in an enclosure. If a photographer was taking a picture of one animal, I kept my eye on all the others. At the same time I keep an eye on the animal the photographer is shooting to (hopefully) predict if the animal is stressed into a dangerous action. The animals' enclosures had good spots for shots, and holes cut in the fencing to accommodate the large lenses professionals have. The animals, however, are not performance trained, walking wherever they chose, not always where the photographer wanted them. This, of course, complicates the job at hand.

I explain all these factors to the photographers before we go behind the guard fencing. It's my job to keep the photographer safe, while they do their work. While the photographer gets into position to take the picture, I hold onto their belts or pants top, ready to pull them back if an animal charges the fence. We have had that happen for multiple reasons. I also explain to the photographer that it is more important for me to save them from harm, than for me to worry about their camera. Photographers are very dedicated to their cameras, so most hold onto them a little tighter while working at the zoo.

A number of photographers and I have become friends—it's hard not to when you've had your hand on their pants! Most of the photographers have appreciated my efforts to keep them safe, even if their cameras got dirty. The male tiger and lion will spray urine to mark territory. Karen Talasco, a photographer who has become a close friend, has taken great photos of animals here at the zoo. Having a photographer close to the fence is a threat to territory. Karen has a camera bag that has been a target of territorial marking. She's a good sport and considers the stinky camera bag a trophy.

It is so much easier to take pictures of the youngsters in the nursery. I have taken raccoons, opossums, fishers, lion and tiger cubs out to roam the zoo for photographers after hours. The natural shots they have taken of the cubs by trees or rocks are great. The rule of thumb is don't take anything out if you can't get it back. Lenny Rue III is the only photographer allowed to take certain animals out without a member of the family being present. He lives by the rules, and has enough animal experience to anticipate how an animal will react to any situation.

I had a little problem with that rule when Kelly Hill, a friend and a photographer from the local New Jersey Herald newspaper, called and wanted a good shot of a young fawn. The Herald was doing a public service piece on fawns saying not to pick them up, momma doe is nearby. Kelly and I selected the smallest fawn, Peanut, from the nursery and took her out of the zoo grounds to a Space Farms' field for a "natural" shot.

Relying on the fawn's natural instinct to lay low and hide in the grass, Kelly and I set up the shot. Peanut lay still and settled into the tall grass of our hay field. Kelly took pictures. Then, rocket scientist that I am not, I suggested we take Peanut to the wooded section nearby for a picture of a fawn in the woods. Kelly liked that idea, so I picked up Peanut, carrying her to the woods. Did I mention that Kelly was five months pregnant? And I had a bum knee that summer?

Peanut had been raised for her seven days on soft, pliable hay. I gently placed Peanut on a pile of old leaves by a log. The sound of the leaves crunching underneath her scared Peanut. Peanut took off deep into the woods. Kelly and I started running, then stopped, remembering that a fawn will instinctually hide. Kelly and I watched Peanut to see where she would settle. Peanut sat down. With Kelly pointing the way, I went to retrieve the fawn. Again as I approached Peanut, she heard me and took off. Kelly and I watched, and lost track of the ricocheting fawn. We searched and searched for the fawn in the section of the woods that she had run. We couldn't find her. Peanut's protective dappled coloring was performing its camouflaging duty. After a half-hour of intense searching, Kelly spotted the fawn by a log, under bushes. I snuck up on the tiny doe and snatched her up in my arms. After all that exercise, Kelly and I both thought the field picture was just perfect. So did her editor—it made the front page.

Chapter 55
What's in a Name?

Names are important. A name gives a person a label for his or her identity. My given name is Lorelee Jean; however, I am not the southern belle type, so I go by my nickname, Lori. Johnny Cash's song, "A Boy Named Sue," shows the importance of a name. A name is the choice of the parents, and I am the adoptive parent of many different species.

Can you imagine being the mom to so many litters of different species and coming up with names for all of them? This is often my problem. I've named animals after characters in the Bible (Gramma always said it was good luck), weather conditions, colors, flowers, people whom the animal reminded me of, fruits, vegetables, natural objects, Native Americans, nursery rhymes, and the individual characteristics of an animal. My Dad was never fond of my naming every animal. It is a communication shortcut. If I had a problem with "Pumpkin," my helpers would know it was the llama with the pumpkin colored rump. The little old lady who lived in a shoe must have had my problem with names. Somehow, inspiration hits you and an appropriate name pops into your head, usually in my case, with a chuckle.

The first calf I raised at the zoo I named Norman after the calf in the movie *City Slickers*. My brother's son, Hunter, was a year and a half old by the time I got the second calf. Hunter learned to say, "Mooo" while we fed the second calf. And Hunter could say "Norman." So that calf got the same name.

The calves we raised got silly names. One Holstein (black and white) calf was named "Spot." Another was named "Rover." The calves came when you called them for the bottle, so it seemed funny to name them dog names. I would go out to feed them, calling, "Here, Rover." You could imagine the look on the kids' faces when a one hundred and fifty pound black and white spotted calf came running.

The first llama I raised was a boy, Jimmy the llama in pajamas. I had to wait a whole year to get a female cria. I had a name all picked out, "Dolly."

Dolly llama was born in the spring of 1998. I loved that name, and its religious overtones.

When the calf came in that spring, I couldn't stop myself from naming him "Elvis." Every day I would announce feeding times over the loud speaker system and kids would come running to help. I always had more kids than I had animal babies that they could help with. New Jersey State law prevents visitors to the zoo from coming in direct contact with exotic animals (any animal not naturally native to New Jersey). The kids were allowed to help with what are considered domestic or farm animals. Llamas are now farm animals, along with chickens, cows, pigs, horses, goats, and believe it or not, ostrich and emus.

The kids were excited to see the animals come running. The parents would laugh when I announced, "Elvis is spotted at Space Farms with Dolly Llama! And it is time to feed them."

One little farm goat came in for my care. Pure white and skinny, she was one of twins and was failing. I was examining her in the office. Jenny, the secretary, asked what I was going to name it. I lifted up her tail to see what sex she was and was surprised by a big glob of yucky mustard yellow poop. "Whoa, Nellie!" I exclaimed. Jenny burst out laughing. "Well, she has a name now!"

Chapter 56
Tara and Khyber, the Tigers

Many interesting human characters have paraded within the history of Space Farms Zoo and Museum. Just as doctors know other doctors, and lawyers know other lawyers, the Space Family knows many animal people. Animal people come in all shapes, sizes and colors, just like the species they deal with. All animal people have one thing in common, the concern, compassion and investment of time and money for their animals. I have seen the spectrum of animal caregivers, the loving circus trainers that live night and day with their charges, zookeepers who truly care, flamboyant shysters that use animals for publicity and city workers who take a zoo job for the winter to get out of the weather. There are good and bad in any profession.

One man who is outstanding in his field is Albert, "Al," Rix. Al Rix was born in Hamburg Germany in 1920. Al was 13 when he learned to train bears in the Stellingen Zoo in West Germany. In World War II Al was drafted into the German army and was assigned to the Russian front. He was shot and captured, returned to Germany and met his wife. After the war Al went back to work with the bears. His ability was spotted by one of the Ringling brothers and Al came to the US with the circus in 1950. Al traveled the globe with his trained bear act becoming a renowned expert in bear breeding and training. After his retirement form the circus in 1962, Al started a bear refuge for his collection of bears: polars, grizzlies, browns, blacks, Hokkaidoes, and Himalayans. The Big Bear Mountain is located in Middletown, New York, forty-five minutes from Space Farms. I have known Al all my life. I have always known him as a petite grandfather type man, with life stories that are enhanced by his heavy German accent. I've seen pictures of him in his younger years. He was a dashing man. Space Farms shares our overflow of venison with him and he shares his overflow of bread (from a bakery near him) with us. He also gave us numerous bears, once the bears needed to retire from his act. Al came to the farm once a month or so, to swap foods, stories and sometimes just to sit and visit. His knowledge of bears was unfathomable. Al once told me he could tell if a bear was trainable (able to learn) by the shape of its head. He is so good at what he does, I believe him. His daughters and son help him at his bear ranch. He has bred endangered species of bears and nursed sick bears at other zoos back to life. Every time he comes, I try to tap a little more knowledge from him. Al just chuckles and tells another road story.

Al had made a deal with a trainer to supply the trainer with two bears. The trainer couldn't pay with cash. Dad and Al were sitting on the front porch gabbing (it's called reminiscing when two old duffers do it). I saw Al had come to visit so I grabbed a soda to join them. Both men had twinkles in their eyes when I approached, so I knew something was going on.

"You veady to be a new mama?" Al asked in his German accent.

"Sure she is," replied my dad.

Hmm... I thought, *we don't need any more bears, and Al's daughters, Jeannette and Susan, raise his. Something is up. Guess I'll bite.*

"Whatcha got, Al?" I asked, knowing these two guys can be jokers.

"I made a trade for zuo tigver cubs and I can't keep zhem, so your dad zays you vill take zhem." Al's eyes gleamed, he knew he was giving me the experience of a lifetime. Dad knew it, too. They both knew I was incapable of turning it down.

"Sure, Al. They can't be too much different than lions, right?" I said. I still could not believe my luck.

"Zoon as zhey are born, I vill let you know," Al explained. My hopes dashed, so many things could happen and I might not get the cubs.

"I'll be ready," I said half-heartedly. Al hobbled to his truck, his character not dampened by his age. Al was seventy-nine years old, a true spirit of the earth.

A month later, ten o'clock at night on October 13th, I got a phone call from Al at my home. Al was in Indiana and on his way home. He would see me tomorrow with the tiger cubs. *Holy mackerel, this is really happening!* I thought. I told Doug and Jackie, then went to bed, but did not sleep, my mind racing with the things I needed to do to set up the nursery for two four-week-old Bengal tiger cubs.

Until this century the tiger was at the absolute top of the food chain, a predator unmatched in the natural kingdom. Prior to 1900 it is estimated that there were hundreds of thousands of tigers, eight subspecies (or geological races) in all. Sizes, colors and markings distinguished the differences. The Caspian, Javan and Balinese tigers are now extinct, the Chinese tiger is down to approximately two hundred left in the wild and the Corbetts, Sumatran, Siberian and Bengal tigers (like the others) are endangered in the wild. With protection programs, the numbers of wild tigers is slowly improving with approximately nine thousand now in the wild. The Bengals have doubled their population in the past few decades and the Siberian tiger population has grown from thirty big cats left in the wild to an estimated one thousand. National and international protection programs such as "Project Tiger" in India and the World Wildlife Fund's task forces are to be given credit for the tigers' possible escape from extinction.

The Bengal tiger is from India and averages ten feet in length. It has the characteristic stripe pattern we are familiar with. The Siberian tiger is the largest of the big cats, averaging ten to thirteen feet long. The Siberian's coat is longer and silkier than his southern relatives. The warm coat and furry paws help to warm him in the cold, Russian winters. The stripe patterns on an individual cat are like a fingerprint, no two the same. The white chin and belly are characteristic of all tigers, while the fur color may range from fawn to rusty red. Striping may vary from black to brown. All of the white tigers in captivity are genetic mutations and have been fostered by creative inbreeding. With human intervention, tigers are born the spectrum of white to pale cinnamon colors, some even having blue eyes. These mutations often have other physical abnormalities.

The tiger is known to be a nocturnal hunter. Scientists have discovered that in protected reserves tigers also hunt in the daytime and concluded that the encroaching human population has caused the elusive hunter to become nocturnal. The tiger stalks his prey and pounces to the neck, killing with its crushing jaws and weight (four hundred to seven hundred pounds). If the neck is not captured on first pounce, the Achilles tendon in the legs will be grabbed until the neck can be broken. The tiger is a killing machine and eons of instinct has perfected the process. Trainers know you don't turn your back on them, even in play a tiger can inadvertently kill the soft humans.

Like all cats, the tiger has great night vision and a good sense of smell. Special organs located in the roof of the mouth register a scent in the air, through a process called the Flehman response. This Flehman response is exhibited as a huge open, gaping, mouth movement that appears as if the tiger is trying to wipe a taste out of the roof of the mouth with the back of its tongue. Males use this technique to locate females in season. The keenest sense a tiger has is his super sensitive hearing. A tiger is able to distinguish the noises of different species.

Tigers are solitary, though siblings may stay together to hunt for a few years after leaving mom. The only contact tigers have with other tigers in the wild is during mating witch lasts about one week. Then the male will move onto the next available female in season. Tiger breeding is vocal and often accompanied with an amount of physical sparring.

Tigers are territorial, the size of the territory dependant upon available food sources. All tigers mark territory with urine, the male's territory overlapping that of several females. Should a strange male, or a young male just proving himself, encroach another's territory, fights will take place, the winner gaining the females favor. Females are sexually mature at three years old.

Tiger cubs are born after a one hundred and five day gestation. Up to six cubs may be born in a litter, only one or two may survive to adulthood in the wild. Tigers born in captivity have a better survival rate. The cubs are born at two to three pounds on average. Again, the smaller the litter, the larger birth weight of the cubs. Young tigers will nurse from momma until they are six to eight months old. Cubs stay with the momma for up to two years in the wild, during which time she teaches them to perfect their natural hunting instincts. The father has no participation with his family.

Man-eating tigers are usually old or injured and unable to catch their regular prey. Other tiger attacks are from man's intrusion into a protective mother tiger's domain. The cases of man-eating are the exception rather than the rule. Tigers prefer deer, camel, buffalo, elephants and occasionally monkeys and birds. Tigers often get themselves in trouble when they prey on cattle, not from the cattle, but from the rancher. These big cats will eat forty to ninety pounds at a sitting. The tiger's digestive system has evolved, like other big cats, for feast or famine. Tigers are lucky if they kill once every two weeks in the wild. After the kill, tigers gorge themselves on the prey, saving the leftovers until they are hungry again. Tigers only hunt when they are hungry, often going weeks between successful kills. Tigers in captivity have life a lot easier. Food, shelter, and medical care are provided to them. The populations of captive generic (crossbred) tigers have been taken off the endangered species list. There are more captive-bred tigers in captivity in the United States alone than there are tigers in the wild.

Al pulled in to the farm just before lunch, on a crisp autumn day. I met him (me, looking out the window every two minutes?) in front of the office where Al, with reverence to his friendship, experience and age, was allowed to park. Al saw me approach and flashed his amazing smile.

"Ere zou go!" Al winked and handed me a small cat carrier he had sitting on the front seat with him. I peeked through the front grate to see two tiny tigers, each the size of a loaf of bread. Four curious clear eyes peered at me, stopping their meowing, cooing voices temporarily. "Zey are verry hungery, I did not veed them, zvo many hands ist not goot." Al followed me into the office with the official paperwork required to bring animals across state lines.

Al gave me a run down on what formula they had been drinking for the three days after they had been taken from mom. Most big cat owners feed their bottle babies on diluted Esbilac. We had had such success with the Zoologic formula 42/25 on lions; I planned to switch the cubs over to that formula. I don't know why that formula works so well for us when other trainers preferred the Esbilac, must be our Artesian mineral water. There is more than one way to *feed* a cat. I mixed up some Esbilac straight away. I would keep the cubs on their established formula for a couple of days, until the stress of transport and new environment had passed, then gradually switch them.

Two bottles, color-coded blue and pink, warmed to slightly above body temperature and I was ready to begin my passage into tiger mommyhood. On my knees on the office/nursery floor, I opened the cat carrier. The cubs that were meowing the moment before shushed and backed to the rear of the crate. I reached in the crate, my trembling hand was greeted with half hearted snarls. The cubs were responding to my different smell, and were somewhat alarmed. I gently grabbed one cub and pulled it out of the crate, giving it a brisk rub down, while cooing to it. Upon being presented with the bottle, the tiny tiger's instinct took over and he sucked the five ounces dry. While the male cub was drinking the female cub heard the sucking noises and bolted out of the still open crate, crawling over me with claws extended. As soon as the male cub was finished, I grabbed the pink bottle for the aggressive little girl. She drank rapidly also, leaving striations of blood on my forearms with her claws. After being satiated with formula I gave the cubs another rub down, stimulated for urine and poop, gave them a quick physical (surprise! they had their carnassial teeth) and settled the duo down in a nest box. Four-week-old cubs are fed every four hours. I left them to sleep the bottle off and sought out Dad and Al who were having coffee in the restaurant.

"Zhey are beautiful cubs, are zhey not?" Al asked. We all agreed. Dad inquired how they drank. I told him they downed the bottles readily. Dad saw the marks on my forearms.

"You better find some gloves, till they get to know you," he remarked. "There are some welding gloves in the stock room." I washed my arms, and then was off to find the gloves.

The cubs were hungry right on schedule. I fed the second feeding with the welding gloves Dad suggested. The cats ate well, the gloves were too short and clumsy, and I could not grasp the bottles with the man-sized fingers of the gloves. Until the cubs caught on to my feeding style, and would keep their feet on the floor, I needed something else to protect my feeding arm. It is natural instinct for cubs to paw the mother's breast to stimulate more milk production. All infants do it, even humans. I found a large piece of buckskin in the stock room and I made myself an elbow length mitten for my feeding arm. It would take a week for me to convince the pair of razor sharp clawed cubs that there was lots of milk for each of them and they need not fight for it. "No Claws!" was the command I used, as I cooed gentle baby talk to the pair.

At night the cubs came home with us, with daughter Jackie enjoying the playtime. Doug was proud of his new tiger daddy status—though he is much too macho to admit to it. I bonded quickly to the tigers and they to me within a few days. These cubs were going to stay with us, at the zoo. I could feel free to love them all I wanted. It is harder to bond with an animal if you have to protect your heart from that eventual separation you know will come. If you know an animal is leaving for another zoo or sanctuary the intenseness of the bonding is stunted. You still love the animal, you have to, or the animal will just not do as well. Just like human infants need affection, so do animals. But somehow knowing the tigers would stay made them more endearing.

I had not picked out names yet; the zoo was going to run a name the tigers contest in the spring. We referred to them as the "babies" or the cubs. One of my photographer friends, Jayne Hindes Bidaut came to visit and asked if she could bring a friend of hers, John, to photograph the cubs. John had done photographic work with tigers in the wild and wanted to see the cubs I was raising. During John's visit he helped me pick out some temporary names: Khyber for the Indian king that a mountain pass was named for, and Tara, meaning brilliant star in Hindu. Names from India seemed appropriate for Bengal tiger cubs, even if the cubs had never seen India and were now frolicking on New Jersey grass. John also told me that the cubs looked great. "Better than I've seen in the wild." I was the proud mamma. But of course a captive tiger cub looks healthier than the animals in the wild. Captive animals have a better, more consistently nutritious food source, supplied shelter and a medical plan. But I was still proud to hear John say the words that Dad had told me many times.

The word got out that we had new tiger cubs, and the local newspapers came to take their pictures. My dad, my immediate family and I were the only ones allowed to touch the cubs. Michael O'Sullivan is a photographer friend who volunteered to take pictures of the cubs on a monthly basis so we could keep track of their progressive growth. When they first came, the cubs were twelve inches long, not including a tail.

Hunter and Lindsey, my brother Parker's children, played with the small cubs under supervision. Lindsey was two years old at the time, and this was the first time I realized she had inherited the family "touch" with animals. She was at ease with the cubs, and the cubs knew it. They were comfortable with her. Lindsey was my assistant whenever she was around, if I was feeding the cubs. She was two years old and did not hesitate to pick up a tiger cub one third her weight and as long as Lindsey was tall. The baby cub would go limp, instinctually. Lindsey would haul a cub across the room to me for a feeding. One time the cubs were squabbling, Lindsey walked calmly over to the cubs, corralled one in a headlock and dragged the cub away from the other, saying, "No fighting" in her tiny two-year-old voice. That moment was poignant for me. Lindsey has the family touch. I decided I would try to help her develop her gift whenever I was able. She has been my helper ever since.

The babies had received mother's milk so their immunity levels were on par with normal. I made an appointment with Dr. Sevideo from Dr. Ted Spinks' Animal Hospital, for their shots and check up at six weeks old. I knew the cubs should be established on their new formula and environment before we stressed out their immunity systems with an immunization shot.

Tara and Khyber traveled with me from home to work daily. Not a long commute, I would put them in an animal carrier in the back seat, fastening a safety seat belt around the crate. At home I had set up a nursery for them on the front porch. My front porch is enclosed just like a room, but it is cooler out there. Since the cubs had come to me in October at four weeks old, and were growing faster than the lion cubs I had raised, I knew that I would not be able to keep them in the house till spring. Dad has always advised, "Don't raise any hot house babies. They may do great while in the house, but eventually when the curtains come down, they have to have the freedom to romp and play. That means outside." The temperature inside a house does not stimulate the animals to grow the secondary fur coat they naturally need to live outside. So their time at home was special.

By the time Tara and Khyber were six weeks old, they had increased their bottle intake to a full bottle (10 ounces) six times a day. I had also started to hand grind venison for them, which they readily devoured. I started them off at five weeks with small amounts increasing the amounts gradually. If you give a baby too much right off the bat, even though they may love the meat, they will get sick, just like a human baby.

Vic Campbell's Cable TV show came with us to Dr. Servideo for the tiger cubs' check up and immunization shots. The cubs were on TV and I'm sure the cubs were thrilled. They did love to watch TV, especially the animal programs.

While at the zoo in the daytime, the cubs came outside to romp and play in the afternoons. The zoo closes on October 31st every year so we can get ready for winter, so there were no visitors around. The cubs loved everybody, they would run and jump on me, Doug, Jack, Dad and any photographers we had visiting. Tara and Khyber were both trained with fingers in the mouth so they were used to being gentle with their mouths (teeth). Running around the zoo was great fun for the cubs (and me). The cubs were very curious, sniffing, tasting and pawing at everything. Jumping, rough housing, and climbing trees (I would pull them down when they were getting to high, it would be embarrassing to call the fire department to get a tiger out of a tree!) filled the hours we were outside.

Tara and Khyber grew at a fantastic rate, quicker than the lions I had raised. By two months old the cubs were two feet long, not including their tails. Playful and healthy, the dynamic duo was getting to be a problem in my house. Evenings they would charge through the house, mouthing everything, including our legs, trying to entice us to play. Our housecat, Daphne, would challenge the cubs to a race around the house. Eventually, Daphne would tire out and jump up onto the dining room table. Tara and Khyber would stop in their tracks with a "which way did she go?" look on their faces. The tiger cubs fell for that one every time. Daphne would gaze down from the table, temporarily confident in her own intelligence. I pinned the curtains up on the rods, not wanting to face the inevitable. While playing with Doug one night, Khyber ripped a hole in Doug's favorite easy chair. Doug hollered, and Khyber tucked tail and ran out to the front porch to his and Tara's nighttime cage.

Every year for Christmas, my daughter Jackie has asked for a puppy. Our lives are so full of animals and we are rarely home, so to have a dog would be unfair to the dog. Dogs were not allowed in the zoo (the zoo animals are afraid of dogs). Also, every year for Christmas I took a picture of Jackie with an animal and had it put on those wonderfully hokey Christmas cards. This year's Christmas card was going to be a gem. I had Jackie put on her red plaid pajamas. We got out the Christmas wrapping paper and spread it around like it was Christmas morning. I got my camera ready and brought the tiger cubs in from the front porch. What I thought was going to be an easy set up shot, turned into a fiasco. The cubs loved the bright colorful paper and tore into it, sending vivid ripped snowflakes into the air. The cubs chased the paper pieces through the house. Khyber developed a new game, putting his front paws on a sheet of wrapping paper on the carpet and sliding it along the floor with his back feet. Both cubs jumped off of the couch and chairs to dive bomb onto the piles of multi-hued Christmas paper, sending more paper flying. The pandemonium lasted for a half-hour, when the tenacious tigers finally tired. Panting, the cubs wandered back to Jackie; Tara picked a particular flat piece of wrapping paper then squatted and peed.

Jackie was scared during the beginning snowstorm of pretty Christmas paper and striped, bounding tigers. The pictures taken reflected her fear and trepidation. When Tara peed on the paper, Jackie was disgusted and finished with the photo opportunity. I convinced Jackie to put one cub in a Christmas bag in front of her. Jackie was fed up and the picture was a grumpy one (of Jackie). Animal photography is very difficult. You have to use what you can get. The picture for the annual card showed Jackie, obstinate faced, and two cubs, one in the decorative Christmas bag with the caption: "But Santa, I wanted a puppy!" When God gives you lemons... make lemonade.

The weather had started to get colder when the cubs were eight weeks old. Two, two and a half feet long rambunctious cats in the house were making the nights more challenging than my days. Khyber and Tara had grown enough to climb out of the four-foot-tall, transportable dog run I had set up on the porch with blankets, pillows and toys. At day break every morning they would climb out to come looking for their morning bottles and me. Suddenly, I would be in bed with two hungry tiger cubs licking my face or sucking on my fingers. I loved the cats, but it was time for them to go outside before the weather got any colder if they were to adjust to the temperature and grow winter fur. I knew the tiger cubs would each be the size of full-grown German Shepard dogs by the time they were six months old, and much more rambunctious.

Dad, Doug and I discussed where to put the tiger cubs for the winter. We cleaned out the old milking parlor in the cow barn. The room was enclosed and insulated. The room had large glass windows that would catch the morning sun to warm up quickly. Doug and Dad built a room-sized cage next to the huge windows so my babies could look outside. On one end of the cage a large, wooden transport box with a double slide was placed. The concrete floor was covered with wood shavings. In the box I placed blankets, pillows and toys from my house. Every day for a week the cubs were placed in the cozy cage. Every night I swore the cubs would stay out the next night. Every night they came home.

My husband, daughter and I go on vacation the week of Thanksgiving. The babies had to be adjusted to the barn by then. The weeks were ticking by. Finally I decided to leave the cubs for the night, taking down their bottles at 9 p.m. When I walked up to the windows, I shined my trusty flashlight into the room from the outside of the windows first. Two cubs were sleeping soundly in the transport box, surrounded by pillows, blankets and stuffed toys. I woke them up to give them their bottles. I should have let them sleep. Khyber and Tara had had a nice nap, full tummies and wanted to play. I played for a few moments and left the cubs to go home. As I pulled away in my car, I glanced back to see striped baby faces with glowing eyes looking out the window. My heart was breaking, but it had to be. Later that night I was tossing and turning in bed next to Doug. Doug knew what my problem was. Finally, with understanding in his voice, he suggested I go check on the tigers. "Don't wake them up, I'm sure they are sleeping, but you won't sleep 'till you know. Go on, you are keeping me awake!" In my jammies, I jumped into the car with my flashlight. Doug was right, the cubs were snuggled asleep with their toys, like ET hiding amongst the stuffed animals. I knew they would be. I knew that. I really did.

Khyber and Tara stayed in the barn for the next week, coming out to play but sleeping the night in their rumpus room. The first nights are always the toughest on me. The animals do just fine.

Jackie was studying Washington DC in school. It was her turn to pick a vacation spot so we were going to DC to see the sights. I made Dad and Parker a list of things to do for the cubs. It was so unnecessary for me to make a list, they knew as much as I did about taking care of the cubs. Dad patiently went over my list with me and promised to continue the 9 p.m. bottles. Dad is usually in bed by 8 p.m. but he kept his word. My stepmom, Mira, would help mix the formula. Parker would feed the chunks of meat the cubs needed. The cubs were still very handleable and loving, easy to care for.

My little family left for DC, I called home nightly just in case. After a week away I was antsy to get home. Immediately upon our return Doug and Jackie were left in the dusty driveway at the house as I drove down to the barn to see my babies. You would not believe how much they had grown in a week!

Back from vacation, we all settled into our routines of school and work. I was at the zoo every morning by eight a.m. with bottles for the cubs. I would play with them, clean their cage litter section, and return to my other work across the street. The cubs were down to four bottles a day, and up to a hindquarter of venison. I brought the tiger pair out to the zoo to romp and play, until snow fell that year. It was getting increasingly harder to get them to behave within the human range of safety. They did not play mean or rough, Tara and Khyber played tigerly, often with teeth and claws. My goal was to raise healthy specimens, not trained tigers. My cubs would grow up free of training, they would be zoo tigers, who get to do as they please within their enclosure. I watched and marveled at their growth. And boy did they grow!

Husband Doug would keep Tara and Khyber busy after dinner while I did the dinner dishes.

Chapter 57
Happy Holidays

Hi everyone, it's that time of year
for Lori's letter of holiday cheer.
Let's see what happened in the year past,
Oh yeah, Doug spent three months on his ----!
His leg has healed, it was a bad break
thank God that's over, just for my sake!
Fishers, opossum, lions and owls
kept my nursery full of growls.
A Holstein calf named Elvis, & by golly
a little girl llama named Dolly.
So Elvis was spotted at Space Farms with Dolly llama!
Speeches and talk at the local campgrounds,
the turnpike, with the Governor, boy I get around.
Sometimes I talk till I'm blue in the face
all to publicize the farm known as Space.
Pictures in the paper, some close up front page,
My God that's not me, I cant be that age!
And on TV - puts on twenty pounds you know,
and wrinkles - oh well it's my twinkle that shows.
More articles published on things at the farm
I type till I get a twitch in my arm.
Enough about me, here's about Jack:
she's so grown up, I want my little girl back.
She's discovered the phone, it grows out of her ear,
might as well get her a headset next year.
One day a debutante, the next an imp
it was a lot easier raising a chimp.
Honor roll student, tall, strong and proud,
still keeps us all laughing out loud.

LORI SPACE DAY

She's the light of my life, gives the world her spark
but then preteens... some days she's a fart.
And Doug, oh Doug, still my sweet honey,
the strong silent type with a great streak of funny.
My little family is doing just great,
Oh and we added two, just as of late.
Two little tigers came into our lives,
with hissing teeth and claws like knives.
My arms look like a bad suicide attempt
Oh well the rest of me looks farm style unkempt.
For vacation we went to Washington DC.
(Who'd look twice at Monica, if they first glimpsed at me?)
Did the sights, the museums, shopping and the zoo
That's what I call a busman's holiday, don't you?
Our other vacation was in Ontario
went fishing and fishing and fishing, ya know.
I don't fish much, I like trashy books,
but I buy a license - just for looks.
Alas, this season is over, the zoo is quiet
so now I'm painting up a riot,
signs, billboards, nothing is safe.
Too bad I can't paint out the wrinkles on my face.
My family's religious high holidays are here,
all the men folk are out hunting deer.
Doug and I don't hunt, but we get more bucks,
bringing back road kill in the death mobile truck!
So there's my life, now how about you?
Write if you can, tell me how you do.
A postcard, a phone call, and email for me,
then I wont' feel cloistered and quite so lonely.
Or better yet come visit, I'd love to see your smile
and we could gab for a little while.
My excuses for this letter, well, 'tis the season
A lot like me, all rhyme - no reason.
Love to all, Lori, Doug and Jackie

Chapter 58
Lizzie the River Otter

On April 4th, 1999 Space Farms had a blessed event. Three little river otter pups were born to our mated pair. We had had a litter two years before and lost the young. Dad asked me if I was ready to raise the threesome, I said yes, but I had some research to do on otter milk. The mama otter was taking care of the babies, so they would stay with her for a while. River otters rarely give birth in captivity. Other zoos wanted Space Farms babies to expand their gene pools.

There are 20 recognized subspecies of otter. The otters at Space Farms are North American River otters. The otter is the most fun loving of all the Mustelidae family, perhaps of the entire animal kingdom. It is often observed splashing and playing with abandon along rivers, streams and lakes of North America. In the harsh winter season, the cold does not affect the otter. Its dense fur and fatty body insulates the river otter from the changing elements. Otters have been seen creating and using repeatedly, snow and mudslides along riverbanks, lakes and streams. The otter has a number of special attributes that make it a remarkable animal. All four feet of the otter are webbed, giving it more swimming power. Each toe does have a non-retractable claw, and the front feet are very dexterous. The serpentine movements of the otter are attributed to its exceptionally slinky spine. Each vertebra is specially constructed for extra mobility, a wonder of nature. This flexible spine enables the otter to "bound" on land in a hump back gait. It also helps the otter in its fantastic swimming and diving abilities. The otter's tail is round at the base and flattened toward the end. Unlike the beavers pan-shaped flat tail, the otter's tail is completely furred. The otter's nose has valves that close the nose while it swims. The tiny teddy bear ears lie flat against the head while swimming and also have valves to prevent water from entering the ear itself. The fur of the otter is so dense and luxurious that it is the standard by which all other furs are judged. The density of the under fur (thousands of hairs per square inch) keeps water from touching the skin itself. This is another insulating factor that helps keep the otter's body temperature constant even in icy, frigid water. The color of the otter's fur ranges from dark brown to black, with a tan throat and underbelly. Otters preen like cats, to keep the fur clean and in place. The musk glands of the otter give it a certain odor—strong, but not terribly unpleasant.

Otters are mainly carnivores, feeding on fish, frogs, salamanders, crustaceans, other aquatic mammals and fowl. Occasionally otters do eat some vegetable matter. I discovered their favorites are carrots and blueberries. At Space Farms the otters are fed on a high protein diet of ground venison and commercial ferret pellets. The enclosure at Space Farms is located in a natural pond, the otters catching their own fish and small turtles much to the delight of visitors.

A full-grown male will weigh 15 to 20 pounds and can be 45 to 55 inches in length. One third of that length is the tail. Females average one third smaller than the males. Usually loners, the otter will travel covering a large circle in a week to ten days. A male otter will cover a territory of twenty miles when food is plentiful. Otters are tracked by scat (poop) and trails. The otter often use the same spot to defecate repeatedly in their travels as a way of marking their territory.

Females will stay near their dens. The otters do not make their own den, but take over an abandoned beaver lodge or muskrat den, or making due under logs in a river bank. Females will not allow males to enter the den. The father may or may not participate in the rearing of his progeny. That just depends upon the male. Some males stick around to help out after the young learn to swim, some do not. The exact gestation of the otter is hard to estimate. Female otters have a delayed implantation of the fertilized egg. The otters at Space Farms have been seen breeding June through August in their first heat. After the first litter is born, breeding takes place immediately (post partum conception). The young are born in this part of the country in April, in other southern areas as early as February.

At birth the pups are blind, toothless and covered in black silky fur. Babies are about the length of a soda can and have the cutest tiny tail. Mama otter's milk is 62% fat and 29% protein of the 38% solids. (The other 62% is water.) This high fat content makes the babies round and pudgy, helping to insulate them against the cold. Mom stays with her young to keep them warm and dry until they "fur out" at five weeks old. At that time their eyes are also open and the teeth are coming in. Otter teeth are nasty and sharp, able to skewer fish easily. At seven to eight weeks the pups are pushed out of the nest and mom teaches them how to swim. Around eight weeks old the otters are weaned. At eight weeks the young are one half the length they will be as adults. River otters are found throughout most of North America. Protected in New Jersey, they are trapped with a special license in certain areas. The otter is experiencing a comeback in northern New Jersey. In the wild otters live fifteen years or so. Like most other animals, captive otters tend to live longer.

 Dad and I checked on the pups every week to monitor their progress. At five weeks old, Dad said it was time to take them away from mama to start the bottle training process. The three pups were adorable pudgy balls of fur. The male was the size of a size nine woman's sneaker, the two girls somewhat smaller. Their newly opened eyes quietly observed all around them. I put the babies in a cardboard box with a heating pad covered by a terry cloth towel. I had contacted an experienced otter rehabilitator in Florida who explained that humidity was very important to young otters. Made sense, otters are born and raised around water. I placed a warm wet wash cloth in the nest box with the three pups. I started them on Zoologic 35/55 formula, which was used by the rehaber, and suggested by the vet. The high fat content (55%) caused the formula to be as thick as pudding at room temperature. Warmed up to body temperature, the milk was thinned to heavy cream consistency. I cut the nipples on the baby bottle in a cross cut to allow maximum flow of the thick formula. The smallest female sucked well, the male and other female took time to adjust to the new nipple. Within two weeks the pups were doing well on the bottle. I offered Gerber baby food in chicken flavor (a favorite according to the rehaber) and the youngsters made a big mess, lapping it out of a dish. But they ate, and that was what was important. Mama otter would chew and regurgitate food for her young—well, there's only so much I am willing to do.

 The pups adjusted to their new surroundings well, and were very loving and friendly to me. We were bonding nicely. Nightly they would come home with me for the round-the-clock care they needed. The otters were off limits for Doug and Jackie. They did not have their shots yet (the otters) and who knows what human colds and/or viruses the otters could catch. After two weeks we lightened up their quarantine and we had otters slithering all over the house. Otter scat is stinky, due to the high protein content of their diets, and somewhat musky. The pups were so cute; I overlooked the yucky part of the job, and the smell in my house. I was training the pups to scat outside. It appeared to be working.

Things were going so well with the otters that we decided they could graduate from the office nursery to the nursery outside in the zoo. The weather was mild and warm by mid-May, so the otters were placed outside in the nursery in a grassy enclosure with a kiddy pool. The threesome played in the four inches of water I allowed them to have in the pool. I had not taught them to swim yet and did not want to take any chances on the pups drowning in too much water.

A week later tragedy struck. The larger male and female skipped a bottle in the morning and seemed somewhat lethargic. The smallest female otter was still sucking well on the bottle. When the larger two skipped the second bottle and seemed warmer than usual, I called the vet's office and made an appointment for early the next day. Dad and I discussed all the things that I should do. I was already doing them, I had quarantined the sick, separated them, kept them warm, administered fluids, and gave them rubs to encourage drinking. The same afternoon the larger female passed away while I was eye dropping sugar water into her mouth. The male was failing also. It all happened so quickly, I called the vet back and rushed the two remaining pups to the office. Dr. Servideo gave them some antibiotic shots and I came home to wait for the medication to take effect.

It was a tough night. The smallest female seemed to slow down but was still drinking. The male was still hanging in there, with fluids administered every two hours by IV. I called the rehaber in Florida to see if she had any advice. She informed me that the immune system of the otter is very delicate, sounded like they got something, and that we were doing all we could. Darn germs again.

Dad signaled me from across the street where he was working the next morning, with the two fingers up code. I gave him the one finger up and one finger iffy sign. The male otter pup died before noon the next day and I took him to the vet for a necropsy. I still had the last little otter.

I spent the rest of the day monitoring the last female otter pup. I prayed every time I fed her, as she was slowing down on the bottle and I was worried. I prayed to God, I prayed to Grampa and I prayed to Gramma. At one point, I was so upset I cried out loud, "Come on, Lizzie, (speaking to my Gramma), help me out!" The little otter, which had only dosed off in my arms, woke up and finished the bottle, eagerly. I named the last little otter, Lizzie, after Gramma, who had also raised an otter in her younger days. Gramma had told me stories of her otter and how much she loved it. I named the otter Lizzie, figuring I was giving her a guardian angel and kept on praying.

THE ZOOKEEPER'S DAUGHTER

The necropsy report came back from the vet. The otters had died from Clostridium Perfrenges, a type of tetanus. Clostridium Perfrenges is found in the grass/dirt of most farms that have goats. Humans are vaccinated against this with the shots they receive as youngsters. Clostridium Perfrenges is a germ the otters were not immune to naturally. Lizzie got a tetanus shot, and all the other babies in the nursery did, too. My shots were up to date. *God, I hate germs.*

I had moved Lizzie and the other two otters back into the office nursery on sterile toweling when they first took ill. I kept Lizzie in the office nursery for a week longer. Lizzie was moved back to the zoo nursery and placed in a wire bottom cage, I would not put her back on the grass again. I would take her out for walks on the lawns outside the nursery. She was back to her normal self, more affectionate to me, as her siblings had passed. She enjoyed her baby bathtub full of water, and the constant flow of visitors that she would "talk" to.

Lizzie became a doll baby. She wanted to be carried all over, just to be with me. When I wasn't holding her, she sat on my feet. She was sweet and affectionate, everything that Gramma had told me that her otter had been. And did I mention adorable? Lizzie looked like a baby seal. Lizzie's dark chocolate silky fur was topped with silver whiskers around her wiggly, shiny nose. Her intelligence shown through her beady black eyes. Inquisitive, cuddly, and constantly chipper, her joyous voice would call to me throughout the zoo, whenever she heard my voice.

Lizzie still came home with me at night. She was no bother at home, learned to scat outside in the tall grass. Lizzie drank her bottles lying on her back in our arms like a baby. She held her own bottles, sucked on our toes, and nipped at our ankles when she wanted to play. Daphne, our housecat, even learned to tolerate pesky Lizzie bounding after her. Jackie played fetch with Lizzie and Lizzie would curl up on Doug's lap when she got tired. Lizzie was fascinated with the dishwasher, she loved the smells (I occasionally let her have table scraps), and played in any dripping water. Lizzie was given baths in our bathtub, in water not over her head. By eight weeks old I figured she should be given the opportunity to learn to swim. I invited local kids to assist in the process. I took Lizzie and met the kids from my 4-H group at Ayers's pond, which was located on Space Farms property. I wanted the boys with me in case Lizzie decided to swim across the pond, or enjoyed herself so much I might have trouble getting her back. I figured teenage boys could swim faster than I could, in case inexperienced Lizzie did, too.

Lizzie was on the shore as I waded out with two intrepid boys into the cool spring water. Lizzie chippered, whistled and cooed to me, wanting to be closer than the five feet away I had waded. My friend Karen Martin stayed on shore with Lizzie. Lizzie raced back and forth along the shore, then decided to take a leap into the water. She sprang from the shore like Mark Spitz at the Olympics, skimming the surface. Lizzie doggie paddled out to me and clung to my legs. Everyone cheered. I picked Lizzie up, gave her some words of encouragement and placed her back in the water two feet from my body. She immediately doggie paddled right back to me, still keeping her head and nose above water. This was a far cry from the diving, twisting, swirling parent otters at the zoo or the otters I had seen on wildlife programming. She needed more distance and practice (I hoped). I carried Lizzie back to Karen who held the squirmy wet otter pup (what are friends for?) till the boys and I waded out farther into the clear country pond.

Ten feet from shore and waist high in the water, I called to Lizzie. Karen put her down near the water. Lizzie dove into the water with delight. Lizzie obviously liked this new game. As soon as Lizzie hit the deeper water she submerged, still swimming straight for me. I stayed in position and let Lizzie find me. "This time," Karen advised, "don't pick her up, she has to learn." Was Karen politely hinting I was being over protective? I let the petulant otter cling to me, then told Ben, Karen's son, to call her. Ben whistled and Lizzie swam to him. Lizzie swam underwater with a push off on my legs like a trained Olympian swimmer. I was thrilled. We all cheered. Lizzie frolicked on and in the clear water for a half hour. Tiring, she went to shore by Karen. I figured that was enough for one day. We went back to the pond many times after that; Lizzie played, dove, turned and swam underwater, learning more from repeated instinct than from the stationary humans relishing in her glee. I truly believe we did not teach the young otter anything, but just gave her the opportunity to become what she was destined to be, a swimming expert of the natural world.

Lizzie was doing great, eating chicken heartily, some venison, fresh caught minnows and carrot and blueberry treats. By July she was off the bottle, but I used the bottle as a treat when I took Lizzie out for speeches. Lizzie was a hit wherever I went, adorable, affectionate and her personality won everyone over. I spoke on otters first that year on the lecture circuit. If I didn't let Lizzie out of her carrying case, she would make loud chirpy noises until I did. As it was people in the audience wanted to know what type of bird I had in the case. People were amazed that it was an otter. I would let Lizzie out, carrying her so everyone could see her. When I was done speaking on otters, Lizzie would play at my feet with her favorite balls and toys. She might wander toward the audience, but as soon as a stranger approached, Lizzie would scamper back to sit on my feet. I learned to stand still to give lectures that year.

August rolled around and Doug, Jackie and I were scheduled for a vacation in Canada at our favorite fishing place, the KO Lodge in Ontario. Lizzie could not come. Dad and I discussed all of the babies still in the nursery and what needed to be done for them in order for me to leave. Dad is the most trustworthy babysitter, after all, he taught me what to do, like his mother taught him. I dreaded where this conversation was going. I knew we had other zoos waiting for otters. You can imagine the joy in my heart when Dad suggested we put Lizzie back in with her parents.

"You mean forever?" I asked in trepidation.

Dad smiled and stated, "Well, our otters are getting older, maybe we should keep her."

"Yeah, that's a good idea," my voice quavered, giving away my joy, while I was trying to retain some professionalism. Lizzie would stay at Space Farms and grow old with us. Hallelujah!

Dad and I discussed the procedure we would use to reintroduce Lizzie to her parents. Not knowing how the parents would respond to the intruder Lizzie, or how Lizzie would respond to her possibly unknown parents, we decided a divided partition would prevent possible fighting and let them get to know each other by sight and smell again, first. The parents were separated into the water section of the enclosure with the wire bridge as a dry area for them. Lizzie was put into the land section of the otter enclosure, with a wire partition between the two parents and Lizzie. The parents were very interested in Lizzie, and mama came right up to the wire to otter talk to her. Lizzie was stunned. She didn't know quite what to do. Mama chippered and called in a nice tone for her long lost baby. Dad otter responded with ambivalence. Lizzie kept coming close to me. My dad suggested I walk away, so with trust in my heart that my Dad knew what he was doing, I went inside the main building. Binoculars in hand, I watched the reunion. Lizzie went over to her mama and they sniffed, bobbed heads and chatted. My dad came in and told me what I already knew, that everything was going fine. Lizzie should stay out over night. It was a long night. I didn't dare go get Lizzie, as I was afraid she might squirm away from me in the night and end up in the pond outside the otter enclosure. I did check on her with my trusty flashlight. She was sleeping, curled up next to her mom, separated by wire, and doing fine. But it was still a long night.

The next day we opened the divider and the mama charged at Lizzie with loving nuzzles. Lizzie seemed a bit overwhelmed to me, but Dad said she was fine. By-mid afternoon the three reunited otters were swimming and cavorting in the water section of the enclosure. All was well. I still fed Lizzie chicken and treats, going in daily to visit and play with her. The otter parents would scamper away while Lizzie and I played on dry land. After a week, we were able to go away on vacation, me with a light heart. Dad would make sure Lizzie got her special chicken diet that she loved. We would gradually switch her over to the venison and ferret food diet of the parents.

Our vacation was wonderful, a nice break from routine, special family time for the three of us. Our lives are so busy, so involved with the daily workings of the zoo from dawn to dusk; like other working families our vacation time is precious, limited to two weeks a year. We fished, swam, read books, played board games, ate great gourmet food, and slept late 'till seven a.m. every morning! What lazy bums we are on vacation.

I checked on Lizzie as soon as we got home to the zoo. She and all my other charges in the nursery had done just fine. Lizzie and the other babies were happy to see me, every one chippering, squawking; llamas moaned their hellos, and welcomed me home in their own way. Work at the zoo continued on its daily pace. By September the nursery had emptied out, native wild animal babies are done needing special care by fall, and the teenage animals were replaced with the adults or moved to larger quarters.

I visited my sister, Renee, in York, PA that fall. While I was gone for three days, Lizzie escaped. Her lighter weight body enabled her to climb over the 45-degree overhang that her parents never could. Upon my return, I checked with dad on my remaining nursery inhabitants. Dad told me Lizzie was out in the pond. Dad and Parker could hear her chirp, she would respond to them but would not come to them. I immediately walked the acre pond's edge, calling to Lizzie. I thought I heard her chirpy noises, but did not see her. I called 'till I was hoarse. I was back at the pond at dusk every night for a week, always thinking that I heard Lizzie. I was concerned, but not overly worried, she could catch fish in the pond and there were plenty of nice spots for an otter to rest. Predators were not a concern by the pond, I knew Lizzie could out-swim even the snapping turtles that lurked beneath the water's reflective surface.

A week later, Lizzie came up to Parker as he was feeding the parent otters.

Parker scooped Lizzie up in his arms. Parker told me he put her back in with the adults, after she gave him wet otter kisses! "Yuck, otter lips," he joked. Parker and Doug had covered the top of the open-air enclosure with light wire to prevent Lizzie from escaping again.

My reunion with Lizzie was a squeaky, otter kiss filled, wet experience. When I heard Lizzie was back, I went out to see her. She was swimming with her parents. I called to her and she came bounding to me in her distinctive otter humpbacked gait, chippering and making the otter noises that had become music to my ears. Lizzie started to climb up my jeans and I cuddled her in my arms. She was soaking wet, drizzled with sand, smelled like an otter, was shedding her summer baby fur, but I didn't care. We played for half an hour as her parents watched with inquisitive looks on their faces. Lizzie nipped my fingers as I rolled her over on her back to inspect her for any possible injuries that she may have acquired on her excursion adventure. The little otter had fared well, did not have a scratch, and had obviously eaten the cornucopia of available fresh pond foods. Lizzie and her parents enjoyed the crisp fall water, swimming and playing most of the day away. Hurricane Floyd came and went, the otters reveling in the higher water to scamper in. Of course, more water would not bother an otter!

A week before Christmas, Lizzie escaped again. We still do not know how. She had happily stayed the three months with her parents. I was back to walking the pond's rim calling her name again. I heard her, but could not see her dark body against the murky winter water skimmed with ice. I was not worried, I figured she'd come back again, just like she did before. She could survive on her instincts and fresh food in the pond. I wasn't worried, but I still walked around the pond every night at dusk, thinking she would want to come home to sleep. She didn't.

Christmas Eve, I was home doing the typical Christmas cooking thing, as Doug came through the yard to the house. I had heard his truck, and glanced out the window to see my husband trudging to the door. I knew something was wrong by his stooped, downtrodden appearance. As I opened the door, Doug threw his arms around me in a giant bear hug, almost a physical restraint hold. As he held me, Doug said, "We found Lizzie."

"Great!" I squealed, "I'm gonna go see her, watch the potatoes don't burn." I tried to get out of the hug to get my coat. Doug held me fast.

"She's dead, honey, she climbed into the Mountain lions and they chewed her up," he said. I cried. There was nothing else I could do. My little otter's vibrant spirit had left this earth, to frolic amongst the streams and rivers above the clouds.

That night was Christmas Eve, my stepmother Mira's big celebration. The entire family would be there. Dad opened the door when he saw me carrying the scalloped potatoes that were my assignment. I put the potatoes down on the decorative, cheerful Christmas table and Dad gave me a big bear hug. "Doug tell you?" Dad knew Doug had, by my red puffy face. It was a tough Christmas. The tears still flow as I am writing this, years later. I have to stop now; I can no longer see the page.

Niece Lindsey and Lizzie Otter enjoy a cuddle.
Photo credit: Mike O'Sullivan

Chapter 59
Chip and Dale, the Flying Squirrels

The Space family has a long-established reputation for taking in animal babies in distress, raising them and releasing them on the Space farm. Only 100 acres are devoted to the zoo and museum complex, the other 350 acres are farmland, swamps and woods. The 350 acres of open space land is a wonderland of environments in which native animals thrive. Our reputation with neighbors and farmers brought me two tiny flying squirrels.

A neighbor owns a local tree grooming company. As his men were out felling a tree for a contract, out ran a mother flying squirrel. After they realized what had happened, the men searched the tree for the nest. Cuddled safe within an abandoned woodpecker hole, and cushioned by the soft makings of the mother's fur lined nest, were two tiny flying squirrels. One of the men called me to see if I would take them in. I asked how big they were and if their eyes were open yet.

"Tinier than my pinkie finger, and no the eyes are not opened yet," was the reply.

I knew that mom was long gone, disturbed by the chain saws and commotion that a logging crew creates. "Sure, I'll give them a try," I said. It was the response the men knew I would give.

"Be right there," the owner of the company replied. "We're working right up the hill." Five minutes later his blue truck pulled into the office.

Sure enough, the little squirrels were as long as my pinkie finger. Their bodies were nicely rounded. Maybe as thick as my other fingers, their mom had done a good job. Soft gray fur covered their backs, the tiny bellies were white, and I did see a few lice. Although their eyes were still partially sealed shut, the infants were aware of the change in their surroundings. Tiny paws had minuscule claws that clung to my hand, in effect wrapping their whole body around one finger. *God, these are small*, I thought. I made them a bed of soft tissues (for diapers), a piece of polar fleece, and rounded up one of Gramma's Mason jars for a hot water bottle. These important items were placed in one end of a shoebox. After dusting the infants with louse powder, I placed the little squirrels in the shoebox.

    I used premature human infant formula for the babies and eye dropped one drop at a time onto their tiny, nearly invisible lips. You know you are getting older when you need your glasses for this type of stuff! The little gray noses wiggled and they took turns sucking in one drop at a time. The tinier the baby, the bigger the challenge. The highly concentrated milk agreed with the babies. I fed the infant flying squirrels every two hours, but they only took a drop or two at the most. I must admit I did not get up in the night to feed them, they did not squeak loud enough to wake me. I bounded out of bed the next morning realizing that I had slept through the night. Luckily they did just fine with me skipping the two and four a.m. feedings.

    Jackie thought the squirrels were cute and we kept the shoebox in her bathroom at night with the door closed for safety (remember we have Daphne, our mouse catching, lion fearing house cat). Chip and Dale were named after one of Jack's favorite cartoons.

    Flying squirrels are nocturnal and small as adults, only measuring about eight inches long including the tail, in this section of the country. This is why few people are aware of the tiny rodents until a flying squirrel makes a nest in their house's attic. Flying squirrels use abandoned woodpecker nests, abandoned leaf nests from other squirrels, or hollowed out trees in the wild. With the expansion of the suburban sprawl in our neck of the woods, many local people have found flying squirrel nests in attics, mailboxes, soffits, garages and even decorative wooden birdhouses.

    Flying squirrels do not fly like birds. They are a glider. The flying squirrel has folds of skin extending from the body proper to its forearms and hind legs. A flying squirrel at rest looks like a small furry gray bat with the gliding folds of skin clinging to the body. The arboreal life of the flying squirrel gives it the height necessary to hop off the tree, spread its folded skin flaps and glide to the next spot. When a flying squirrel extends its folded skin to glide, it becomes a sail. By tucking the legs one way or another, the flying squirrel works the air similar to the rigging of a parachute. The shaggy tail of the flying squirrel acts as the tail on a kite does, a counter weight to the heavy head, helping the squirrel to direct his glide. Diminutive in size, the flying squirrel's small body weight also helps in the gliding process. An adult will weigh only 2 ounces, so you can imagine how small the little ones are.

Large, protruding pea-sized eyes give the nocturnal flying squirrel excellent night vision. Tasty prey for hawks, owls and snakes, the little rodent must be constantly aware of his surroundings. Flying squirrels, like all their squirrel cousins, are fast and skittish. Flying squirrels are no shirks on land; they can scamper as fast as their gray squirrel cousins can. Raising them from small babies, the flying squirrel does become an interesting pet. However it is currently illegal to do so without specific licenses. Flying squirrels range in the United States east of the Mississippi River, from Maine to eastern Texas. Favorite habitats are wooded and swampland areas with trees.

Young are born in the spring after a forty-day gestation. The momma squirrel nurses her babies for eight weeks at the end of which time the young are two-thirds the size of mom. Like most rodents the squirrels need to chew constantly or their teeth will overgrow and cause a problem. Flying squirrels like to feast on seeds, nuts, berries and fungi and are one of the only squirrels to eat baby birds and insects.

The tiny squirrels became a part of our family, traveling in their shoebox. As they got larger, the two traveled with me in a Longaberger basket that had a lid. Daytimes were spent in a vertical fish tank equipped with branches for climbing. After they were officially off the bottle, the petite pair was fed on an assortment of grains, rabbit pellets and their favorite treat, Honey Nut Cheerios. This favorite kids' cereal is also a favorite of most rodents. I've used them to begin other small rodents on solid food.

Chip and Dale went with me to the campgrounds and were a guaranteed audience pleaser. I would select a volunteer from the crowd, usually a small boy and girl, and tell them I was going to teach them a new dance, "Like the Macerena." I would place the flying squirrels on their t-shirts and tell the kids to catch them. The roving rocket rodents would race up and down all around the kids' bodies. The kids would twist and wiggle, flailing their arms, giggling while the crowd laughed. The squirrels were searching for the Cheerios I kept in my pockets, but the kids did not have any. A few of the kids would catch on, that to catch a flying squirrel you have to stand still and hold one hand high. The squirrel instinctually seeks the highest point. I did however hold the true secret. Whenever I wanted the squirrels to come back to me, I held the sweet nut cereal in my hand, then gently closed my hand over the palm-sized young to place them back in the Longaberger basket.

I had learned from previous experience not to release tame squirrels in the trees at the zoo. Tame squirrels tend to freak out the city folk, when they come jumping onto the picnic table to enjoy lunch with them. Chip and Dale were released in the woods by the railroad tie bridge. I watched them scurry up a tree, sniffing and tasting their new surroundings. I visited them for a while, but when they would no longer come to me when I called, I knew they were wild squirrels once again. That was fine, since that's what they were meant to be.

Chapter 60
Baby Beaver

Beavers are a fascinating animal and well adapted to their environment. Beavers have been dated to prehistoric times. Prehistoric beaver skeletons were found in Indiana, the live body size estimated to be 700 to 800 pounds, according to Dr. Leonard Rue III's book, *Furbearing Animals of North America*. The much smaller modern beaver played an important part in the early exploration and settlement of the North American continent. The fur of the beaver fueled the trappers and explorers westward in pre colonial days. The mountain men and trappers searched the continent for the beaver, whose fur was the fashion of the early 1800s. Beaver hats were the rage in Europe. The European beaver (Castor fiber) was trapped to near extinction in the search for more pelts for hats, coats and robes. The price of a beaver pelt drove the trappers and mountain men to move farther and farther west, trading with Native Americans along the way. Soon after, silk replaced the beaver as the fashion fad, but not before the beaver was practically wiped out east of the Mississippi and the countryside had been thoroughly mapped out. The beaver's comeback, ranging throughout the North American continent, is a testament to its industriousness (Busy as a Beaver?) and protective building practices.

Beavers are the largest North American rodents, like a rat or a woodchuck, and have two huge chisel-like incisors. Their main diet is the bark and leaves of soft woods, aspen, willow, poplar or birch. The beaver is one of natures more conservative creatures. Beavers eat the leaves and bark of trees, then use the leftovers, the remaining uneaten wood, to build lodges and dams. The construction techniques of the beaver are so intricate and strong that man must use dynamite to destroy a beaver dam or lodge.

Beavers (usually a pair) move into an area with a stream and rapidly cut down the closest surrounding trees to build a lodge and a dam. The dam floods an area enabling the beaver to swim to the nearby trees. Beavers are excellent swimmers. However, they are clumsy and slow on land, making them easy prey.

The flooding of an area gives the beavers safe and easy access to more trees. A beaver can fell a twelve-inch diameter tree in a few hours and smaller trees take much less time. Doug, Jackie and I were camping one night in Allegheny National Forest and a beaver felled a tree not twenty feet from our tent. All we heard was the whoosh as the tree crashed into the water. It was the next morning before we realized it could have crashed onto our tent as we were sleeping.

The lodge of the beaver is its home. Lodges are usually round in shape—a dome of dead branches and twigs sprouting from the water. Always built surrounded by water, the beaver's lodge has an underwater opening. The underwater opening prevents predators from entering the den, ensuring the sleeping beaver's safety. The inside of the lodge is lined with uneaten branches, leaves, and twigs in the late fall. All winter long the beaver eats his "wallpaper." The beaver also stores trees and larger branches underwater for winter forage. With this foresight, the beaver does not have to come out on the frozen pond and waddle to trees for food. Beavers are very vulnerable on land, as they are cumbersome and slow walkers. While the beaver may taste nasty to humans, a hungry coyote, fox or wolf is not so picky.

Lodges are multi-layered. The first floor of the lodge is only a few inches above the surface of the water. This room is used for eating and drip-drying from the swim in. The second level is used for sleeping and rearing young. The beavers chew wood down to shavings to make the nest comfy cozy.

The beaver has a round compact body. An average adult beaver will weigh 40 to 60 pounds. The females are as large as the males. While the front feet are very dexterous, only the hind feet are webbed for swimming. Beavers are known for their tail. The beaver's tail is a leathery, flat paddle. The skin of the tail is crosshatched with lines, giving it a scaly appearance. The tail is used as a rudder while swimming and as a balancing post while sitting up to gnaw down a tree. The tail is also a communication device between beavers. Beavers will slap the water surface with their tail when danger is near. The loud splash/slap on the water notifies other beavers in the area that a predator is nearby. The tail is not used for carrying building supplies as seen in many cartoons.

Both the male and the female beaver have castor glands (oil glands) near the base of the tail. Castor is a smelly oil used for grooming, communication and marking territory. Beavers regularly preen themselves, spreading the oil from their oil glands over the fur, waterproofing it. This enables the beaver to keep his skin dry as the thick, wet fur compacts during swimming. The oil helps to preserve the fur and makes it silky to the touch. Beavers come in many color ranges from a soft dishwater blonde to a deep black brown often with an auburn tinge as they age. Beavers in the northern parts of the continent have the darkest fur. The fur has two layers, a soft one-inch long undercoat and a layer of two-inch guard hair.

Beavers have special adaptations that help them survive in harsh winter environments. This includes an enlarged lung capacity to help it stay underwater for up to fifteen minutes. The eyes are equipped with a second clear eyelid (nictitating membrane) that protects the eye and helps them see underwater. When the beaver goes underwater its internal body functions slow down to help conserve the oxygen in the blood. A fatty subcutaneous layer and the beaver's fur helps to insulate it from the frigid temperatures. The beavers know a neat trick for swimming under frozen lakes. When the beaver is out swimming and runs out of air, it exhales his breath. The exhaled breath rises through the water collecting oxygen once again. An air pocket is formed under the ice and the beaver sticks his nose into it. By breathing the re-oxygenated air, the beaver expands his underwater swimming range for food.

A beaver colony consists of groups of parents. Each family has parents, pups from that spring and the yearlings from the previous year. Each set of parents creates its own lodge. If the lake or stream is big enough a number of family lodges may be constructed, and a colony is formed. If food is not as plentiful, or the original colony has become overpopulated, the yearlings head out for other streams, rivers, ponds, or lakes, with a little prodding from mom and pop.

Beaver breeding takes place in deep water. Beaver babies are born after a 107-day gestation. Pups are born in May in this part of New Jersey. The pups are born completely furred, eyes open and with orange chisel teeth already protruding though the gums. Pups nurse for eight weeks but start chewing wood for the soft bark at two weeks. Baby Beavers can swim at twelve hours old and take to the water easily without instruction from mom, although they do stay close to her.

My older brother Eric stopped by the zoo early one morning. He had established his own business as a suburban trapper after leaving the family zoo business. Eric's business entailed live trapping native wild animals out of the surrounding suburban homes. He often live trapped squirrels, raccoons, bats, snakes and other critters from the houses of people not accustomed to country life. The local farmers used Eric's business to trap out beaver and muskrats that moved into the farm ponds used for livestock water. This particular day, Eric had three little beavers. He had previously trapped and relocated the mother beaver to another wilderness area. A large man, with a larger heart, Eric was worried that the young would not find their mother, and would I raise them till they could be released on their own? I looked into the box to see three balls of fur the size of muskmelons, plus little beaver tails. *How cute. How could I say no?*

Eric knew I could not resist and I took the three puffballs out of the box for a preliminary physical. Although they had a few ticks, they looked healthy, eyes clear and rectums clean. Beavers have a cloaca, a single opening at the rectum. The intestinal tract and urinary tract empty into the cloaca. Sexual organs (of the male) are contained within this internal section. Long story short, you cannot tell the sex of a beaver by just looking, unless it is a nursing female and the nipples are distended with milk. So the three beavers were named Al, Bert, and Cindy (just in case one was a girl). I carried the box into the office, grabbed Dad and a pair of hemostats and set to work. Dad held the wiggling beavers and I pulled out the ticks. I estimated their age at 6 to 8 weeks, which is near weaning time. I tried them on a bottle a couple of times but gave up when I saw all three beavers chewing on the sapling willow branches I had given them.

The babies were set up in the nursery, in the grassy pen Lizzie was no longer using, as this was during the time she was recuperated and living in the wire bottom cage. The aqua blue kiddy pool was set up for the pups, filled with water and small, tender willow tree branches. The beavers did fine, eating, growing, and swimming the days away. Every morning I would empty and fill the pool, a yucky job. Beavers must defecate in water. Their fibrous diet packs inside the cloaca and is rinsed out in the water naturally by the beaver's body.

One morning the beaver count was down to two. I could not find the third pup. Dad tracked the beaver, and since there was no sign of foul play, we assumed the mini beaver made it into the acre large zoo pond next to the nursery.

The other two beavers, Al and Bert, went on the lecture circuit with Lizzie and myself. My speeches that year were on aquatic mammals. Beavers are non-aggressive animals and were easy to handle due to their small size. The large orange teeth and special tail fascinated the audience. Early September I let the now-teenage beavers go in Ayers's Pond. Ayers's had a large willow tree that had crashed into the pond and was sprouting all over as willows tend to do. The young would have lots of food and the nearby swamp bogs offered protection from the elements. And what ever happened to Cindy? She (?) is still living in the pond at the zoo. She has eaten all the multi-floral rose bushes around the pond, trimmed out the large willow trees, and devoured the tiger lily, daffodil and tulip bulbs it took me hours to plant on the island. She has taken over an old muskrat burrow under a large willow tree and is still there in the pond today. We see her now and then, safely sunning on the opposite shore.

Chapter 61
Khyber and Tara Graduate to the Zoo

Khyber and Tara were still living in the old milking parlor room of the barn across the street from the zoo. All winter long they grew longer and bigger on the sections of venison they were fed. My kitties still played with the balls, stuffed animals and their favorite—long strips of deer hide. When we put the tiny tigers in the barn, they were the size of a full-grown beagle. We knew they would grow fast and large, too large to walk across the road to the zoo on leashes and collars. By May the eight-month-old striped teenagers were larger than German Shepard dogs and able to put their paws on my shoulders, and I am 5'7". Khyber and Tara would greet me with chuffs and kisses. I was constantly saying, "Down." There is just something about having tiger teeth so near to my own that it makes me nervous.

I am a great animal mommy, but I am not a great trainer. I had kept the tiger cubs on bottles way past the time they would have stopped nursing in the wild. This was done for four reasons: #1. They liked the bottles (yes I will admit that I was spoiling them), #2. If they became sick, it is a lot easier to put medicine in a bottle, #3. It was a bonding factor between us and the cubs, and finally #4. We used the bottles to entice the cubs to go where we wanted them (positive reinforcement).

When the cubs had grown large enough and played too rambunctious for humans to interact safely with them, I cut round holes in the back of their sleeping/transport box. I would give Tara and Khyber their bottles through these holes. Then the slide section of the front part of the sleeping/transport box would be dropped, locking the tigers inside. This enabled me to clean their cage without having to watch my back and legs (the cubs loved to practice their hunting and stalking techniques on me). After the cleaning was done, I would let the tigers out of their large, wooden box. This process was developed with long range plans in mind.

In May the tigers were to graduate to the main zoo. Space Farms' public relations man set it up. All the local newspapers, radio stations, and television stations were invited for the big move. May 20, 1999 was to be graduation day, the day the two tiger cubs graduated from the nursery to the zoo proper. The press people arrived right on time. Dad, Parker and myself talked to the press people, explaining the process and giving out information on tigers in general. I was nervous. My little babies that first came to me the size of a loaf of bread, were now big enough to knock me down if they felt so inclined. Their tiny toothpick teeth had grown to be six-inch fangs. Toenails I had clipped with a human baby fingernail clipper were now three inches long. My babies had grown up. One third their future adult size, Khyber and Tara were loving and tender for the first few moments, then they wanted to play. This meant like tigers with tooth and nail. I had not been in with Khyber and Tara for three months, always using the bottles to entice the excitable cats into the sleeping / transport box. Today Parker and I would be inside the large 1/2-acre grassy section of the tiger cage with my overgrown cubs. There was lots of room for the cubs to get a good running start on their favorite instinctual game—catch and kill.

Dad and Doug drove out to position the wooden crate full of felines next to the entrance to the grassy section of the enclosure. The press, Parker and I walked over the hill after Dad called us on the walkie talkies to tell us the tigers were getting antsy in the transport box. As I cleared the ridge of the hill, I saw the box lined up next to the chain link fence. The four-feet by four-feet wooden cube was literally shaking. The cubs wanted out, yesterday. Parker and I entered the grassy section by a side door. I had Khyber and Tara's baby bottles loaded and in my pocket. In one of the zoo trucks was a loaded gun, just in case. Parker and I positioned ourselves for the best photo shots, across from holes cut in the fencing for cameras. The shaking wooden box with the tigers was one hundred feet away. *The cubs will love to romp and run in here*, I thought. *Hope they don't jump on us, I wonder if Parker can hold up under a hundred and fifty pound pounce, I know I can't. Watch my back, Parker, and I'll watch yours. OK God, Your turn.* I handed Parker a bottle.

Parker took the bottle and gave Dad the thumbs up signal. Dad opened the slide door of the wobbling wooden crate. Tara and Khyber stopped their cavorting and lay down. They lay down in the crate, looking at me! The press waited. The cubs stared. I was dumbfounded. I had totally expected the cubs to come running and jumping on Parker and myself, not this. I had raised chicken tigers. The giant cubs were afraid of the press, afraid of the open space and even apparently, afraid of the green grass. Dad whistled and hit the box. The cubs still sat. Dad told me to call the cats. I hollered, "Hey Kitty, Kitty!" just like I always had whenever I greeted the cats. All of a sudden with a leap and a "Rrrrreooooow," Khyber came bounding out. Tara was not to be left behind. She was running right behind him. The two sibling tigers came toward my brother and I at full run. I braced myself for the impact. Khyber came first to Parker with the bottle. Tara came to me, jumping up to put her front paws on my chest. I backed her down with the bottle in my hand. The bottles were emptied lickety split. The useless bottles were thrown over the fence so the tigers did not get angry that the milk had run out. Parker and I patted the cubs' bellies for burps and then gave them a good rub down. After a few moments I knew the cubs would want to play "tigerly" soon. Parker and I decided it was a good time to leave. Dad flung a deer hide on a rope over the fence to distract the terrific twosome on queue. Khyber and Tara took the bait and ran to the hide that Dad was now dangling in the air. Claws extended they ripped into the deer hide. While the tiger twosome were playing tug of war (with teeth and claws), Parker and I snuck out the door of the enclosure. Parker put the padlock on the gate. The press got their pictures and stories and left for their offices. *Whew!* The actual time Parker and I were in with the tigers was maybe twenty minutes but it felt like a lot longer. These were cubs I had raised from four weeks old. Can you imagine coming up against the absolute king of the food chain in the wild?

Chapter 62
Hurricane Floyd

People who work with animals are constantly aware of the weather. We have to be prepared for the various temperatures and the coming weather conditions to keep the animals safe and warm. Space Farms is located in a valley and has a topographical location that seems to attract storms and occasionally tornadoes. During bad thunderstorms my brother, Parker, patrols the zoo to make sure none of the surrounding trees have been hit by lightning, throwing branches or trees across fencing. Tree branches can short out electric fencing. During the storm the animals hunker down, like their wild counterparts do, and wait out the storm. A tree trunk crashing down on a fence can make an animal escape possible. A tornado came through the zoo in 1980, tore the roof off the main building holding priceless museum artifacts and felled a tree across the wolf enclosure fencing. Weather is a constant concern.

Hurricane Floyd came through the northern part of New Jersey on my birthday, September 16, 1999. It knocked the electricity out for miles around, tossed branches, and, in general, made a big mess. The water from the zoo grounds poured into the zoo pond, with mud leaves and water cascading over the barriers, seeking the lowest point. The pond was filled to capacity, but the manmade dykes between the pond, the sika deer and the goat enclosures held. The stream in the goat pen overflowed its banks, leaving the ten goats only a small section of high ground. The otters enjoyed the higher water levels. The last spring's baby beavers in the zoo enclosure swam into the local stream leaving only the two larger beavers in the zoo. The family is always "on call," but during weather storms we all stay alert to the possibilities.

The morning after the storm passed, we were all at the zoo bright and early to fix whatever damage nature had inflicted. We had lucked out, nothing drastic had happened, just a lot of water, mud and branches to clean up, and road surfaces to scrape flat again, that kind of work. I went out to feed Spot, the calf, and the three teenage llamas that were running with him in the long enclosure bordering the zoo's pond. When I saw Spot, I was surprised to see porcupine quills protruding from his muzzle. One of the llamas had quills in its nose, too. I quickly checked the neighboring porcupine enclosure to make sure our zoo porcupines were there. They were. I scanned the trees and the shore for a porcupine, but did not see one. *Humph, one must have washed in,* I thought.

I went back to the building to find some help and the hemostats. I explained to Dad what had happened and he came out with me to de-quill the calf and the llama. If we did not de-quill the animals, the hollow quills would slowly expand and sink deeper into the animal's body, causing infections and other problems. The longer you wait the deeper the quill sinks. Left unattended, quills will embed themselves completely until hitting a bone or an organ with disastrous results.

A half-grown Holstein calf is a strong creature and porcupine quills are imbedded with a hook on the inside. Dad was having trouble holding the squirmy calf. Parker was going by on the tractor with the scraper blade for the roads. He saw we were having trouble and jumped off the tractor to help. With the two men holding the calf, I used the hemostats to start to pluck the quills out of the calf's tender nose. Spot was hurting. Dad held the calf's ears, Parker held Spot's tail and they managed to wedge Spot between the fence and the men's bodies. I was working the hemostats as quickly as I could, yanking the quills with as delicate a tug as necessary. The quills jammed up on my hemostats, so I, in true farmer fashion, wiped the quills off on the back of my jeans. Spot had the quills in his nostrils and had been snorting to try to remove the quills. This made the quills slimy, sticky and gooey. Spot had about fifty quills in his nose so the process took some time.

Next we de-quilled the llama that had come in contact with the porcupine. The llama was easier because of the cush that a llama does. When Parker grabbed the pony-sized llama, it simply gave up and stood there, intimidated by Parker's strength. Dad held the fuzzy chocolate brown head while I pulled out the black and white, two inch, hooked quills. The llama's muzzle was smaller, and only had about twenty-five quills. Every three or four quills, I wiped the quills off the hemostats on my jean-covered fanny, farmer style. The animals were under distress so I was working as fast as I could. I'm sure from a distance I looked like I was doing a 1960s version of the pony dance. Finally, after a half-hour of pulling quills, the job was done. The guys released the llama, and we straightened our aching backs.

We turned to step up and out of the enclosure, when I felt a sharp stabbing pain in my right butt cheek. "Yeow! Yeow!" I hollered and stopped in my tracks. Parker was ahead of me and Dad was behind me. I twisted around to look at my butt. All I saw was Dad laughing so hard I thought he would bust a gusset.

"Seems you got a problem," Dad spit through his laughter, "You want ta drop your jeans, or should I just pull 'em out through your pants?"

"Just pull 'em out," I replied sullenly, as I braced myself against the fence. I assumed the position and Dad quickly pulled out the quills that had worked through my jeans into my skin. I didn't think it was all that funny, but the guys did. The llama, Spot and myself recuperated without a problem. Their noses and my fanny were sore for a couple of days.

Chapter 63
Sweet Pea, Cowboy, Chiquita, Turk, Dale and Annie, the Lions

Sweat Pea and Cowboy were born on September 18th. The momma African lion was not paying them any attention. Matter of fact she didn't even stay in the den after their births. I was a lion mommy again. Nice big babies, Sweat Pea and Cowboy were growing nicely on the formula and routine I had established and followed for lion cubs. Sweat Pea was a lover, a truly affectionate little lion that loved to cuddle. Cowboy was all man, swaggering John Wayne style on bowed baby legs throughout the house. Three weeks later, the second momma lion had her litter of four. Not only was this momma not paying attention to her litter... she was carrying one around dragging the poor thing across the concrete section of the lion habitat. Parker and Doug went in the den to bring out three, two males and a female. By dusk the female had tired of her last toy, leaving the bruised baby in the grass of the habitat. Dad brought that female cub home to me, tooting his car horn in the driveway. He handed the pitiful little thing out through the window of the car to me.

"See what you can do, but don't blame yourself on this one," were Dad's coded words. That pretty much meant "this one is a goner."

Sweat Pea and Cowboy were already established in their boxes, the three new cubs, Turk (the color of the collar I had for him), Chiquita (so blonde like a spotted banana), and Dale (the smallest so far) had drank to my satisfaction all afternoon. They were asleep on their heating pads in their individual boxes. The last little lion Dad brought me was a mess. Jackie was in her room doing homework and I recruited her as an emergency assistant. The cub was covered in dirt and grass from the habitat that clung to her sticky birth fluids the momma hadn't bothered to clean off. The little female was cold, so I started a bath of warm water for her to warm her up quick. I placed her in the kitchen sink. (Doesn't everyone wash their babies in the kitchen sink?) With one hand, I held her lemon-sized head above the water. With the other hand I gave her a one-handed rub down under the warm water. The tiny cub started to wiggle. I gently washed her face and ears with a wash cloth, avoiding the newly forming scab on her nose. Jackie was standing by with dental floss for the umbilical cord and towels for the dry off. After drying off the rat-like cub, I inspected her for more injuries. Her little toes were still encrusted with dirt and grass in the scabs caused by her momma dragging her across the concrete. The little cub started to make sucking noises, which was a good sign. I gave her two tablespoons of warm formula. She drank wearily, and was immediately asleep. I still needed to do something about her feet before infection set in. Antibiotics would surely screw up what little positive bacteria were in her stomach. I still hadn't cooked supper for us humans. I recruited Jackie for help. I had Jackie sit on her bed while watching TV and hold the petite clean cub's yucky front feet in a cup of warm water with Epson salts.

I fixed supper and called Jackie and Doug to eat. Jackie didn't come, so I stepped into her room to see what was going on with the cub. Jackie teared up as I walked into the room.

"Is she going to make it?" The cub was asleep on her lap, paws still soaking in the now dirty water.

"I don't know, sweetie, she's had a hard luck life so far," I replied with a sigh. "She's still breathing, Pop Pop says there's always hope."

"A hard luck life, like Orphan Annie?" Jackie questioned. "That would be a good name for her, Mom." And Annie got her name. Annie thrived, she was a scrapper like her namesake, and held her own with the larger, stronger sisters. Annie had personality plus. Within three weeks the fur grew back on her paws, her nose healed and she looked like the other little lion cubs bouncing around my living room.

The three males had new homes to go to before they were born. Everyone wants Atlas African Lion males because of the magnificent black mane. Cowboy went to a private owner in Pennsylvania, Turk went to Miami Florida, and Dale went to Nevada to a trainer who was going to put him in show business. Two more agonizing trips to the airport. The female cubs are harder to place. Personally I prefer the females. Instinctually they are more social. The pride is made up of females, grandmas, aunts, cousins and sisters. The females hunt for the entire pride, bringing back food for the nursing, geriatric, and sick. The male, the supposed king of the beasts, hunts for himself in the wild, and only participates in pride life when the females... deem it necessary. The male orbits the pride. Young males are chased away from the pride by the head honcho (alpha) male upon their sexual maturity. They form teenage gangs that will eventually split up to fight another head honcho male for their own harems. Sometimes brothers will stay together and rule as a team. The males I've raised have always been loving to me, their adoptive mom, and playful as cubs, but somewhat distant emotionally. Not that I didn't love them just as much as the females.

The female cubs are much more loving, rubbing on your legs, licking and preening my skin, and giving me kisses. But the females don't get those fancy manes that are the showcase of the males. I get very attached to my babies, and winter was approaching fast. I knew I could not keep the lions in my house all winter, due to the size and activity level they would achieve by spring. I could not raise them as hot house babies. Sweat Pea, the gentlest little lion I've ever raised, Chiquita, and little Annie stayed in my house till the second week of November. Our Thanksgiving vacation was coming up and my three girls had to be acclimated to the cooler winter quarters in the old milking parlor by then. My girls would stay in the old baby tiger rumpus room until we figured out what to do with them. I prayed nightly for a good home for them. In the meantime, I had a lot of fun playing with them, taking them out for runs through the now closed zoo. Chickens, beware!

Jackie, Doug and I went on vacation to Pittsburgh to visit our friends there. Upon our return, the mini pride of lionesses had grown larger. They were eating solid food readily. Annie was still the smallest and scrappy, but her paws had healed properly. Sweat Pea was forever the lover, a limp slinky in my arms. Chiquita was a blonde, the lightest lion I've ever seen. And then a phone call came in.

A zoo in Arizona, called "Out of Africa," wanted the two younger cubs. Only two. I tried to convince them to take Sweat Pea saying that she was a sweetie, a loving docile lioness. Well, they weren't sure if they wanted three lions. I explained that they had been raised together and I did not want to split them up and leave one alone. The girls (and I, but I didn't say that part), the girls had bonded. It was two weeks before Christmas. "It's sorta like a Christmas sale... take two get one free?" Ok, they said they'd take Sweat Pea, too. *Yahoo,* I was so happy. My girls would grow up together in the same pride. I made the arrangements at the airport, got the paperwork in order and told Dad he had to drive them to the airport. "I just can't do this trip, Pop." I had gotten too attached. I should be more professional, but it just doesn't work that way for me. Dad understood and I helped pack the girls up for the awful ride to the airport. Dad is tougher than I am and he was not as attached to the girls. The Girls... they did just fine, arriving on time and joyfully playing with their new keepers.

Karen Talasco, my friend and a great photographer, sent "Out of Africa" a disposable camera. *Why didn't I think of that?* Out of Africa sent it back. My girls are doing great, wonderful specimens of a beautiful pride, romping through the Arizona sand. The keepers play with them daily. I do think it is ironic, however, that three little cubs born in Beemerville, NJ are living in Arizona in a place called Out of Africa. The only Africa they have seen was on the Animal Channel on my TV.

Chapter 64
Y2K Bug Holiday Letter

Hello to friends both near and far,
I bet you're wondering how we are.
Here's your answer, my year in a poem
from the lady who lives where the buffalo roam.
Raised lots of babies, much to my delight,
but that part of my job keeps me up nights.
Eight lions, three white tails, an otter named Lizzie,
kept my life more than busy.
Three beavers, ten llamas and two flying squirrels,
kept me in a constant swirl.
I juggle formulas all through the summer,
visitors, speeches and hard labor (bummer!).
Oh, let me announce, my humble self
received two awards now on my shelf.
One for Sussex County Tourism Volunteer,
One NJ Achievement in tourism last year.
Don't get me wrong it sure was nice,
To be recognized not once, but twice.
My 4-H Club , based at the zoo,
is making great strides I'm telling you.
Received a National 4-H Council award
to make a coloring book from a bulletin board
they had created, (and it won a blue ribbon)
on wild black bear safety, I'm not kidding.
So like little elves, twelve kids are busy
sketching a coloring book, the concept easy.
Not anti or pro hunting 'twill be,
then off to print, to distribute to all the kiddies!
Only four grants given out nationwide,

makes a girl's heart swell with pride.
Doug's doing better, his leg's sorta fine,
He still keeps me laughing most of the time.
Jack's a teacher's dream, so they tell me,
a normal teenager's all I want her to be.
Though I really hate going to the mall
it's what teenagers do after all.
Of all my years that I remember,
this has been my worst December.
Went to the podiatrist for a heel spur,
but have three, I found out from her.
On the way home I was in a "bender"
a car cracked into my back fender!
I was not hurt, neither was she,
Just messed up the back of my GMC.
To Lizzie my otter, I'd grown quite fond,
then she escaped and was into the pond.
Not to worry, in a week she was back -
but only stayed long enough for a snack.
Escaped again, this time not so smart,
visited mountain lions, my broken heart.
She wasn't eaten, just played with too rough,
I've cried like a baby, Oh Yeah, I'm tough.
The Grinch stole my Christmas, now what's with bug Y2K?
"Tomorrow, Rhett, tomorrow is another day."
So that's my year, hope yours was better.
If you find time, drop me a letter.
Or better yet come visit me,
A feminist in a redneck community.
Sometimes things just go wrong -
Like a sad country western song.
So cheer me up drop me a line,
I love to hear from friends of mine.
Love, Lori, Doug and Jackie

Chapter 65
Painting in the Zoo

A lot of people have told me they want my job. It is the joy of the animal contact they want, or perhaps the novelty of working with exotic animals? What non-zoo personnel do not realize is the amount of behind the scenes non-animal work that goes on. By March the worst of the winter weather is gone, and spring clean up begins. We have, however, had deep snows as late as mid April, and that sometimes sets us back. As soon as the weather breaks in the spring we are outside cleaning up the zoo after the winter cold. Think about it. It is impossible to clean up manure while it is frozen to the ground all winter. Manure is hauled out in the spring. Uneaten food is also frozen to the ground, along with discarded wet bedding. During the January thaw, which usually occurs in this section of the country, some clean up can be done. This is all back breaking shovel work, and dirty, yucky work to boot. Everyone volunteers to help feed the babies or participate in animal enrichment programs, but no one ever volunteers to pick up a pitchfork. Yes we have tractors, and yes we have machinery, but most of the work is done by hand. The animals don't mind the tractors but the noise of the smaller engines may bother the multi-leveled hearing of an animal. Or the dens are not large enough to bring in a piece of equipment. So each animal enclosure is treated differently. There is an element of danger involved also. Still want to be a zookeeper? Bring your own shovel.

The one hundred acres of grounds that comprise the zoo need to be raked of leaves, twigs and branches that fell over winter. These are shoveled up and hauled away. Loose rocks, heaved up by the frozen ground, must be removed so visitors do not trip. Signage is returned to the zoo. I take the signs inside in the fall, which helps the signs last a few more years. It usually takes me two days to replace all the signs on the enclosures. Some, of course, need to be remade and replaced. Feeding machines that are available to the public are overhauled and put out once again. Roads are scraped flat, after the winter plowing unevenly distributes the dirt and/or pavement.

The Museum buildings are dusted and swept. The main building and the restaurant are thoroughly scrubbed top to bottom, ready for the impromptu board of health inspection that we always pass. The mounted animals in the museum are dusted and sprayed with moth spray, to prevent the obvious problems.

While I have covered this process in a few paragraphs, the actual work time takes a month and a half, seven a.m. to six p.m., six days a week before Space Farms is ready to open in the spring. We have Bill Raab, the farmer and long-time family friend, and usually one other worker, besides family members, working in the zoo. It is a lot of work, tiring and dirty, but at least it is outside. Needless to say if it is raining hard in the spring, that sets us back, and we concentrate on inside jobs for that day, if possible. After the basic cleaning is done for the spring opening, the spruce up work begins. Spruce up means anything that looks shabby needs to be painted or replaced. If it needs to be painted, it's my job. Not that I am any great artist, I just seem to be better (neater?) than anyone else is here. If it needs to be replaced, it's Doug's job. Doug is a mechanical wizard.

I have painted at the zoo all my life. Cement blocks, wire, walls, buildings, swing sets, cast aluminum rocking horse type children's animal rides, you name it I've painted it, probably twice maybe three times during my life. In my Pittsburgh years, I would paint signs and send them home with whichever relative was visiting. When I came home to visit, I would paint cage bottoms.

I do have an artistic flair, which I developed while I was a stay-at-home mom with Jackie in Pittsburgh. So every now and then, after a wall is base painted, I steal the time to paint a mural or some other decorative thing. That is my reward after a long day's work. Doug and Jackie are very understanding. If I am into a project, they have learned not to bug me about supper. Doug will start the grill or get a pizza. He is a prize.

I painted the Little Red School House, climbing to the top of the twenty-foot roof and perching myself, straddling the roof ridge to paint the cupola. I was scared, I don't like heights, but I did it. I've climbed a conveyer belt to paint the top of the corn crib style Macaque enclosure, with a paint roller on a rod. I climbed a ladder resting on the outside of a three-story barn to paint fake windows on the plywood covering the broken windows. Again, no great artistic achievement, black squares on white wood, but I climbed that ladder at 46 years old. Museum buildings may be scary because of the height, but animals are downright dangerous if you are not paying attention.

The Kodiak bear enclosure needed to be painted. It took one day to scrub the concrete and cement blocks clean of accumulated dirt, bits of dried food and shed bear hair. It sounds like an easy job, but keep in mind less than one foot away was a huge bear that wanted to play with the "toy" I had, the hard bristle barn broom. Doug helped me out by weed whacking (the bears did not like the noise of the weed whacker, by the way) the grass close to the ground near the breeding den section of the enclosure. The breeding den section is on concrete for sanitary reasons. Again, we had to watch the weather to ensure three or four days of no rain in order to prepare the surface and paint. When all the prep work had been done, I set out to paint the enclosure.

The Kodiak bears we have were hand raised by Al Rix, so the bears are friendly, or at least the male is. The female shies away from us, but we can't trust either of them. Neither bear is afraid of humans. I wear a large sun hat when I am working in the zoo. With a wide-brimmed hat on my head and comfy work clothes to bend and stretch, I started the paint job. Dark green is my favorite color for zoo bottoms. I was happily rolling paint on the edge of the concrete, sun shining on my back, when the male bear decided to play. He swatted my paint roller. I pulled the paint roller back and waited for him to leave. The bear meandered away. I went back to painting and whoosh! The bear was back in the blink of an eye and had swatted my roller again. This bear was huge, about twelve hundred pounds, maybe ten feet tall standing on his back feet. Suddenly the bear has dark green toenails. I laughed and thought *Better than blood red, especially my blood red.* I realized that the big old goofy male was just playing and was not about to stop. So I moved to another section, squatted and started painting closer to the ground. The base of the concrete is four feet high, immediately above it is a six inch clear area for scraping out the big bear patties, or leftover food. The male was preoccupied with his newly painted green toe nails, (what male wouldn't be?) on the other side of the enclosure. I started to paint.

The sun was out, it was a beautiful day, and I was accomplishing real work, something more tangible than the press releases or public relations work I do. Or my housework that needs to be repeated daily. Not to say that press releases, public relations or keeping house are not work, but for me the satisfaction of accomplishing something tangible, that you can see and say, "I did that," is much more rewarding. Must be that Feminist/Puritan work ethic, if there is such a thing. Anyway... I was happily painting along and kaboom! My fanny hit the grass, my arms akimbo, the paint roller magically appeared ten feet away in the spring grass as my hat danced away in the breeze. Facing me, on the other side of the bars, was the male bear, like an overgrown cub happily at play.

"That was fun," he seemed to be sneering, "let's do it again." I shook myself off and re-captured my hat from the breeze. My straw hat had a gaping hole in it. *Humph, he must of got me.* I checked my head, but found no wound. *Might as well finish the job.* I finished the job without my hat, realizing after that close call that my wide-brimmed hat had impeded my vision. I did not see the bear coming above me while painting below him. My neck was a little stiff the next morning but it was a lesson learned. A lesson learned really cheap.

When the new dens for the big cats were constructed in the spring of 1997, it was my job to paint them. The cougars and jaguars were no problem. Space Farms has one nasty leopard that will grab you as soon as he looks at you. I enlisted a friend to watch my back when I painted close enough that the stealthy, mischievous cat could grab me through the cage with an extended paw and claws. We had to stop several times, paint in a different area and then switch back when the spotted cat was preoccupied somewhere else. I don't believe in taking chances.

I don't mow the grass inside the guard fencing, the guys do. They have the same type of problems with playful animals.

Chapter 66
Lucky the Fawn

Space Farms received a routine call on May 13th to pick up a white tail deer that had been hit by a car on Lusscroft road, which is close to the Space Farms Zoo and Museum complex. My brother Parker took the call and headed out after the supposedly dead dear. When Parker arrived on the scene, he quickly assessed the situation and decided emergency measures needed to be taken. The doe was fatally injured, barely alive. Parker inspected the doe and found she was pregnant. Knowing that the fawn would be close to full term at that time of year, Parker performed an emergency c-section on the side of the road. Parker safely delivered a small, full term, female fawn. Parker removed his t-shirt and wiped the gooey fawn clean, stimulating her to breathe her first breath. "The fawn perked up immediately and tried to nurse on my chin," Parker explained to me later. "Then the fire alarm went off!"

Parker was the fire chief of the Beemerville Volunteer Fire Department at the time. Another local volunteer fireman was on his way to the Beemerville Firehouse, passed Parker on the side of the road, doubled back and picked up Parker and the fawn wrapped in a t-shirt. They radioed ahead and Dad met Parker and the newborn fawn at the Beemerville Firehouse. The Beemerville firehouse is located right across the street from Space Farms and was, at one time, Space Farms land later donated by my family. Chief Parker handed the delicate fawn, wrapped in a Harley t-shirt, off to Dad and jumped in the fire truck to go to the fire. Dad immediately brought the tiny wet fawn to me for the zoo nursery.

"Here's a surprise." Dad smiled as he quickly eased the fawn into my arms, "Parker just delivered it by c-section, see what you can do." With that said, Dad rushed back to his post at the Firehouse. He is the radioman now. Dad has been Chief, as his dad Ralph was before him. Gramp started the Beemerville Fire Department. There are three generations of Fire Chiefs in our proud family.

I stood in the restaurant, slightly agog, at the suddenness of the exchange. I looked at the tiny fawn in my arms and headed for the zoo office nursery. I finished the drying off and cleaning of the newborn. It was a girl. I tied off her umbilical cord with blue dental floss and placed her in the empty incubator. She needed to rest in the incubator's warmth after her traumatic delivery. While the fawn was resting, I prepared the formula she would need. Hoof stock is a little tricky if they have not received any mother's milk. The immunities of the mother are not passed through the placental barrier like carnivores are. These immunities are passed through the first milk and very necessary to the newborn for its future survival. I had a supply of colostrums, the nutrient and antibody-rich first milk from a cow, from a local dairy farmer who kindly supplies us with however much we need. I thawed eight ounces out in a pot of body temperature water. You can't use a microwave or super hot water as it kills the antibodies. The fawn drank greedily on the first try, newborns usually do. The newborn's natural instinct to suck is amazing. The fawn settled down under the heat lamp and slept.

Dad and Parker came back from the fire (a car crash). Dad checked in with me to see the fawn. "We'll have to call her 'Lucky.'" Dad approved of the care I had given. "She is so darn lucky her mom was hit so close to the farm!"

And Lucky she was. Lucky finished her colostrum two ounces at a time every two hours, and was switched over to the Land of Lakes Kid goat milk replacer we preferred to use on fawns. Land of Lakes, yeah, the same folks that make the butter. Lucky progressed and stood for her bottles by the time she was six hours old. Hoof stock are usually up and able to run in a few hours after birth. If not, something is wrong. Lucky's soft hooves dried out during that time, and her umbilical cord dried out also. She stayed in the incubator for a couple of days and then graduated to a cardboard box with hay and a heating pad.

A week later the May weather was warm and balmy. I decided to put Lucky outside in the nursery run set up especially for white tail fawns. Basking in the sun like her wild counterparts, Lucky slept and drank the afternoons away. I went out to feed Lucky one time and could not find her in the fawn run. I looked all over and through the high grass we purposely left for her to hide in. She was nowhere to be found. I went back to the main building to see if Parker's kids, Hunter and Lindsey, would help me find her. I explained to the kids that we were going on a "sort of" Easter egg hunt, only we had to find the fawn. We looked all over the zoo grounds near the run. No Lucky.

Finally, Lindsey cried out, "I found her, Lee Lee!" (Lee Lee is the nickname the kids call me.) Tiny, four-year-old Lindsey crawled under the front porch of the Country Store that sold candy in the zoo. Sure enough, there Lucky was, curled up in a spot only a short, four-year-old could see. Lindsey was proud of her accomplishment, and I was glad to have a new apprentice with a short perspective. Lindsey has been an "Animal Mommy Helper" ever since.

A week later it was to be warm at night so I left Lucky outside in the grassy run, driving down in my jammies at 9 p.m. with her last bottle till morning. The native fawns were outside all the time, I rationalized. Early the next morning I fixed the bottles for the nursery and walked up the hill to feed Lucky first. I spotted Lucky and my heart sank to my toes. Lucky was strewn out in the morning dew, her neck and head stretched backward touching her little spotted back. Lucky's eyes were rolled to the back of her head, her tongue extruded from the side of her mouth. *!#!, This is not good.* I carefully picked up the convulsing fawn, supporting her flailing head and legs. Back to the incubator she went, heat lamp on high. I sought out Dad to see if there was anything else I should be doing. I couldn't think of anything, but Dad has an amazing supply of farmer remedies.

"Heat is the best thing right now, Lucky just got too cold, I've seen calves come back from the brink of death with just a heat lamp," Dad said. I crossed my fingers, called the vet just in case he had any better ideas, and the vet said the same thing. Nothing to do but wait.

I finished my nursery chores, praying as I made my rounds. Two hours later, I checked on Lucky again. She had curled into the proper sitting position, but had not raised her head. A good sign, but she was not out of the woods yet.

Another two hours ticked by. A watched pot never boils and sitting by the incubator with the fawn would do no good. Finally, Lucky raised her head, eyes in the proper position, tongue in her mouth, and nudged my hand when I touched her. I was so happy I did a happy dance in the office. I rushed to the kitchen to prepare a bottle, which Lucky drank like nothing had been wrong. *Whew, that was a close one.*

Lucky was kept inside at night for another month till the end of June when the weather reached 65 degrees at night. She was the only fawn in the nursery that summer. Lucky became quite the celebrity after the newspapers heard her amazing story. Visitors came to the zoo with one question: "Where's Lucky?"

Lucky has stayed with us at the zoo. Imprinted by close human contact, it would be cruel to release her. She is not afraid of humans, or more importantly, cars, trucks and lawnmowers. During hunting season she would be an easy target. Losing her fear of automobiles because of her exposure to the farm vehicles, she would certainly be hit by cars or trucks while crossing the road. So Lucky is still with us, frolicking with the bucks this last fall. This past fall mating season was the only time Lucky was not strolling the fence looking for a friendly pet on the head or a goodie from her human friends. She knows she is a deer now and will most likely have a fawn next spring.

Chapter 67
Bloomfield Boomerang, the Hawk

Every spring the zoo receives dozens of phone calls on what the general public perceives as orphaned, abandoned, or injured wildlife. It is seldom true. Most phone calls are a process of education, with us explaining that the situation is normal, just leave the animal alone, and it will be fine. We also warn the caller of possible exposure problems the animal may cause humans. If we think the situation warrants it, we refer the caller to an animal rehabilitator licensed by the state, trained to help the animal until it can be returned to the wild. In a captive zoo population, you cannot take the chance of introducing a disease from the wild. If we do take in an animal for however short a time, it is done on a case by case basis, and kept in quarantine for an appropriate time period.

I was called to the phone on a young Kestrel that had taken up residence in a church's rafters. The caller was a taxidermist from Bloomfield (a city an hour away from us), and had done some taxidermy work for the family.

"How do you catch a flying hawk from the rafters?" he asked.

My pet shop years really helped me out on this one. I simply replied, "Wet birds cannot fly, shoot him down with a super soaker water cannon, ya know, like the kids use."

The American Kestrel is a protected bird, and cannot be taken from the wild permanently. I thought the taxidermist knew this. I assumed the bird was captured and released when I heard nothing further from the taxidermist. Three weeks later the taxidermist showed up at the zoo with the small hawk in a cage.

"A gift for you," he explained. I wanted nothing to do with this hawk. This was, in my opinion, an illegal bird. I explained this to the taxidermist who just sort of looked at me with that "you don't like my gift" look we've all seen on the mother-in-law's face. I told the man that I would make sure it could capture its own food, then release it. He seemed satisfied with that and left the small brown bird with me. The hawk was capable of capturing its own food. I fed it a live mouse to prove that fact to myself. So I needed to release this bird as soon as possible. I had learned my lesson about releasing birds of prey too close to the zoo's population of free-range chickens, geese, ducks and swans. The zoo still has a problem with magically disappearing chicks. So I needed a place far away to release the Kestrel. I was giving a speech at Camp Taylor in Columbia, NJ that evening. The Taylors were long-time friends of the family. Founder Joe Taylor and my gramp used to trap together in the old days and the next generation, Clayton and Jeannie, are friends of mine.

I called Clayton and explained the situation—could I release the Bloomfield Kestrel at his campground? Of course I could. *Great, one problem off my mind.*

"By the way, Clayton, you don't have any chickens do you?"

"No," he chuckled, knowing my problems with the zoo's chickens. We'd laughed about my chicken problem before. That evening I took the hawk with me to the campground. Camp Taylor is a wilderness campground set in the forest of the Delaware Water Gap. I love the atmosphere there. It is one of the few wilderness campgrounds around. Camp Taylor has the true flavor of the northeastern woods, forested campsites, mossy rocks, a large lake to swim in and is also home to the Lakota Wolf Preserve. The Taylors have done a wonderful job of combining wilderness with the creature comforts (showers and flush potties are big with me). Animal and nature savvy; the Taylors strive to educate their campers to enjoy the great outdoors without harming it. I enjoyed the forty-five-minute ride to the campground deep into the prime New Jersey forest. Clayton greeted me upon my arrival. His rich brown beard, plaid shirt and blue jeans gave him a natural backwoodsman's appearance.

With a twinkle in his eye, he said, "So this is your problem bird? He looks healthy." I unloaded the zoo animals and museum artifacts for my speech. Clayton, ever the gentleman, helped me carry in my things. "The bird has no problem, but I might if I don't release him!" I joked. We laughed and caught up on our families.

I gave my speech on the animals I had brought in the open air raftered pavilion that is located on the campgrounds. I saved the Kestrel hawk for last. I had explained to the audience the pros and cons of animal rehab, how to determine if an animal is ready and capable of being released, the problems

I've had with magically disappearing chicks and why I was releasing the hawk at the campground in the wilderness. With what I thought was great flair and showmanship, I opened the cage of the hawk. I had purposely not handled the hawk. I did not want the hawk to know me. I opened the door... the hawk sat on its perch. I gently shook the cage, the hawk held on tighter to the perch! After my great speech on releasing this hawk the darn thing did not want to go. Clayton's eyes silently laughed at me from the back of the crowded pavilion. Clayton's son and Clayton exchanged a few words that I could not hear in the front of the room. I didn't need to. I saw Clayton's son laugh and slap his thigh. *This laugh is on me, boys.* In truth, I had to hold in a chuckle, too. Good thing the audience was looking at the eight-inch-tall, brown, frightened bird. I slowly removed the sliding grate from the bottom of the cage, lifted the cage into the air and stuck my hand up through the bottom to attempt to scare the striped hawk out through the cage door. The little Kestrel fluttered to the front door, resting a moment. *Come on, go, darn you!* Then he took off. *Finally, Thank God, there are seventy-five people watching, how embarrassing.*

    The kestrel raised in the rafters of a church went home... to the rafters of the pavilion! Not exactly the freedom seeking soaring into the wilderness I had envisioned. It was a good thing Clayton was on his toes as I watched the hawk in embarrassed amazement. Clayton calmly walked over to the electrical box and turned off the ceiling fans. One thing I had not even thought of, *Kestrel chop suey, oh, great! Headlines the next day: Zoologist creates new protected dish.* The laughter and commotion of the crowd had little effect on the church-raised hawk. Neither did clapping or swaying of hands and arms. *I thought it was a Catholic Church, not Pentecostal Baptist.* The small hawk flew a half dozen laps around the pavilion, a boomerang in constant motion. I was still facing the defiant bird of prey totally dumbfounded, when Clayton produced a broom. *Why didn't I think of that? Daaah.* A broom was the eventual instrument of conversion for a hawk raised in a church. After twenty minutes the hawk had been released and re-released and chased from the rafters of the pavilion. Clayton has not seen the small hawk since. But then again, Clayton has no chickens.

Chapter 68
A Soused Boar in New York City

The New Jersey State Fair ®/ Sussex County Farm and Horse Show is held the first week of August every year. It is located in Augusta, which is just five miles from Space Farms. Space Farms sets up a display at the fair and I was a member of the Board of Directors. Donna Traylor is a personal friend of mine, an excellent photographer, and her brother is a meteorologist for *The Today Show* in NYC. Donna arranged for a spot in the crowd outside on the plaza for The New Jersey State Fair ®/ Sussex County Farm and Horse Show . The Fair Mascot, a huge rooster, a Holstein calf and a Space Farms baby Llama, Dolly, were packed up at four in the morning for the hour and a half trip to the Today Show. The Sussex County Agricultural Agent, Dan, his two teenage daughters, my daughter (all 4-Hers), the Horticultural agent, Brian (who wore the rooster suit), an exchange student, Donna, the llama, the calf and myself caravanned into the city.

The parking garage was a block away from the plaza. We arrived at 5:30 in the morning. The truck and the van were unloaded. The calf was leash trained. My little llama was not. Dolly would follow me around the farm so I found no reason to leash train the week-old cria. I expected fuzzy, white Dolly to walk by my side like she did at home. I was, once again, wrong. Little Dolly Llama was afraid of the city noises. I carried her cradled in my arms. The parking lot attendant looked at us in amazement, never cracking a smile. We all made quite the entourage strolling down the city block, as you can imagine. We reached our destination at *The Today Show* and the stage personnel set up people cattle gates to hold the animals in. *It's amazing how similar these people cattle gates are to the real cattle gates. The Today Show* staff were very helpful.

We all took turns going for coffee and watching over the animals. Our teenage 4-H girls enjoyed the sights of the city around the plaza. The show did not start till 7 am so we had quite a bit of time on our hands. While we were waiting for our spotlight moment to promote the fair, the pedestrians crowded around the animals. Not often you see a calf and a llama in the middle of the city, I guess. These animals were everyday sights for us. The crowd came and went on their way to work.

From the edge of the curb I heard a voice. "Oh it looks like country music day in the city." The voice started to sing a country song. The words were slurred. I located the source of the obnoxious voice. It was coming from an obviously intoxicated man. Dan and I looked at each other and grinned.

"Takes all kinds," Dan said.

"Yep" was my laughing reply.

The drunk was scrawny, alcohol or drugs had eaten away his body and mind. This man wove closer to us in the cattle gate corral. Then he spied the three teenage girls with the animals. Dan and I were watching him, my motherly instincts on high alert. *One drunk in a block radius and he has to pester us.* Even though clad in black leather and pretending to be tough, the skinny little drunk posed no threat. Any of our girls could have knocked him over with a feather if they could have gotten past the permeating stale smell of booze and cigarettes. The drunk got louder as he got closer.

"Oh, country girls, won't you come out tonight..."

Dan and I chuckled once again. I guess we are just not used to drunks on the streets. Heck, we don't even have sidewalks at home. I decided to step between the girls and the sight line of the derelict to stop the ogling bloodshot eyes. The sodden man looked at me—one of those looks up and down that every woman detests.

"Oh you are a real country girl," he cooed, in what he may have thought was a seductive tone.

"Yes I am," I replied curtly. He stepped closer to me, forgetting the teenagers that had originally attracted his attention.

"Whoa, mamma, you are a big country girl."

That's it, time to put this drunken pig in his place. I stepped as close as I could to the odiferous offending man. In a low voice, that I thought only he could hear, I calmly stated, "Yes I am. And I know how to castrate pigs without anesthetic."

Dan had heard the interchange and was trying hard to hold in a laugh. I couldn't see Brian's reaction behind the mascot rooster head. The little wasted man put both hands up in a sign of surrender and backed away. The guys and I laughed. There is more than one way to emasculate a soused boar.

The rest of the morning went quickly, the girls in their 4-H t-shirts, the rooster mascot, the calf; llama and Donna were on the weather section. Donna got a nice plug in for The New Jersey State Fair®/ Sussex County Farm and Horse Show. Our promotional trip to the city was a success. My personal reputation as a nice person, however, was somewhat tarnished after I bullied a drunk.

Chapter 69
The Moose and my Caboose

    Every summer Doug, Jackie and I take a hard-earned, week-long vacation. Like every other working woman the preparation to leave work (especially the babies) and home involves a Herculean effort. At the zoo, Dad takes over the nursery with some help from the kitchen staff. Parker and the other zookeepers cover any of Doug's multifarious jobs that can't wait till he gets home. My life is run by my watch, feeding schedules every two hours, group bus arrivals, departures, employee break schedules, meetings and the other management concerns for fifty weeks out of the year. This is one of the weeks I just take my darn watch off.

    Our favorite vacation spot in the summer is the KO Lodge in Deep River Ontario. The KO Lodge is another family run operation, run by the Carlin family. It is on the edge of Canada's Algonquin Park and is a fishing or hunting paradise. Far away from the hustle and bustle of our daily lives, we spend the days fishing, swimming, reading, eating (the food in their Sportsman's Lodge is unbelievably great), and, in general, relaxing. Doug loves to fish. Fishing is a hobby that I, as a country girl, grew up on. We have been fishing together forever. Jackie loves fishing also. Even though she is growing older, she still loves to go to the KO lodge. Could be the hot tub, swimming pool, shopping in town or the boys, but she still enjoys our family vacation. We don't know how long that will last. The KO Lodge is located on Sullivan Lake and has a river edge lodge two miles away, on the Ottawa River. While visiting the lodge, guests have their choice of those two fishing areas or any of the twenty-four nearby Canadian lakes. We tell the Carlins where we want to go and they set us up with a boat and directions. We have fished the Canadian lakes for a number of years. Jackie caught a huge Musky when she was eleven. The Muskie and bass seem to bite the hooks of the Days. The fishing is phenomenal, however I enjoy the scenery and the abundant native wildlife just as much.

LORI SPACE DAY

I have always wanted to see a moose in the wild. Matter of fact I've never seen a moose in captivity either. They are hard to keep in a zoo setting due to their massive size and dietary requirement of aquatic vegetation. As a youngster, I was always jealous of my brothers when they went on hunts out west, Canada, and Alaska with Dad. Not that I've wanted to hunt, but to observe the wilderness and the animals firsthand has always been interesting to me. Dad and the boys always came home with tangible excitement and great sightings, even if that one "got away." Moose are seen all the time in the Deep River section of Ontario. It was my goal from the first year we went to the KO Lodge to see a moose in the wild. Supposedly they are all over the place like the native white tail deer here in New Jersey. Every day on vacation we would get up, leave our cabin, and go to the lodge to eat a big breakfast. Then we would head out on our fishing trip for that day. Doug and I would put the boat, supplied by the lodge, in the crystal clear water of the lake of our choice. It takes two adults to put a bass boat in the water. I am usually the one to get wet. It doesn't matter. Part of my goal on vacation is always to even out my farmer tan. I wear a bathing suit under my shorts. Picnic lunches are part of the package, so we often fish till mid afternoon. After loading the boat back on the trailer we would cruise the back roads of the Algonquin Park looking for a moose for me on our way back to the lodge. For five years! For five years this was our pattern. I was well equipped with binoculars, camera, and antacids (those back roads are bumpy).

Never one to give up on a challenge, Doug, Jackie and I would often ride around sans boat in the afternoons into dusk, looking for my elusive, noble savage moose. Swamps and forested areas are the favorite hang out of the moose. We've seen a lot of Canada's beautiful natural landscapes.

On the last fishing day of our trip in August of 2000, we decided to take a fishing trip to one lake that had a notoriously bad boat launch, Tall Pines Lake. We liked to fish that lake. The rock bass were fast biters and numerous. For me, the trauma of putting the boat in on a rutted, rocky, forty-five-degree launch was always detrimental to my enjoyment of the day. Doug and I got the boat in the water with me only getting my pants wet to the knees. Not so bad. The day was hazy and the wind chilly. Not a day to work on my tan. We had dressed warm in long jeans, t-shirts, and sweatshirts. The fish were biting like crazy. Jackie was pulling rock bass in so fast I could not read more than one paragraph in my trashy romance novel at a time. (I don't always fish.) Fish always seem to bite the best when the weather is the worst. There is no such thing as a fair-weather fisherman.

Doug was happily casting away from his position at the front of the boat when he happened to glance at the sky behind us. Large, rolling, black clouds were moving in. Doug and I decided that it would be best to call it a day and get off the lake. Doug and Jackie reeled in quickly. We prepared to rev up and get out before the thunderstorm hit. We were racing the rain down the five-mile lake. We made the boat launch in record time. I jumped into the water thinking in that spot the water was only knee deep. I was wrong. The water cascaded up to my waist, soaking me to the skin, my jeans weighing me down. Doug ran up the steep launch to get the car. He backed the car down and since I was already soaked, I hooked up the boat to the car. Doug put the Jimmy in four-wheel drive and spun mud as the boat cleared the water. Jackie was instructed to stay down. The rain and lightning were too close for comfort. We quickly unloaded the boat during the spitting rain. Jackie raced me to ride "shotgun." Naturally, she won. I stripped off my soggy, muddy jeans the last thing before I jumped into the backseat of the Jimmy. I wrapped up toga style in a beach towel around my waist, figuring I could sneak into our cabin in my underwear without being spotted. The deluge hit then and we sat in our car till it passed. Afterward the forest looked like the Garden of Eden, freshly washed, misty and sparkling in the sunlight.

We were shaking ourselves off, settling into the ride back to our cabin when Doug suddenly slammed on the brakes. I looked up from my book. There in the middle of the dirt path, fifty feet in front of our Jimmy, was a magnificent male moose. He was wonderful. The moose was obviously a young one, but still larger than a draft horse.

"Got your camera, hon?" Doug quietly said.

"I'm getting it," I replied, as I twisted to the back of the car to get the bag I'd stuffed full of paraphernalia from the boat. Camera in hand, I slowly opened my car door, putting one foot out at a time so as not to scare my sacred moose. I had to get at least one shot. I popped my head above the car door and took a picture of the mighty moose as he stood standing in the middle of the road.

"I'm gonna get closer," I whispered.

"Be careful, he may charge," Doug responded.

I work with animals, too, hon. I cleared the car door and carefully walked closer to the dark chocolate brown velvety moose. The moose glanced in our direction and I froze still in my wet sock tracks. The moose looked toward the swamp across the road. I crept closer. The moose looked toward me again, I stopped, again. I took another shot. The moose twisted his head toward the forest behind him as if he did not know which way to go. His tail switched, and skin twitched, ridding him of the biting flies that were also attacking me. One more look in my direction and the young male moose took two giant steps and disappeared into the swamp.

I had seen my moose! I was thrilled. That is one check off for my life book list of things to see or do. I watched him till he blended into his surroundings as nature had intended. With happiness in my heart and a large grin on my face, I turned to walk back to my waiting family in the car. There, sitting in the front of the car, were Jackie and Doug laughing and cracking up beyond belief. They were pointing at me, gleefully belly laughing at me.

"What's up?" I questioned.

"Look down," Doug shouted. I looked down. My towel was gone. It had fallen off when I stepped out of the car. In my excitement of seeing my moose, I hadn't noticed. *Oh my God*!

I was standing in the middle of the road in my wet, white (now sheer) underwear and a damp t-shirt. I hobbled as fast as I could on the pebbly surface in my wet socks back to the car. I hopped into the backseat as Jackie and Doug were still cracking up.

"I wish I had a camera," Doug spit out between belly laughs. "The moose was sure looking at you funny, I would of called it The Moose and the Caboose." He slapped the steering wheel.

"Good thing no rangers came around, Mom, you'd be arrested for bare shooting out of season!" Jackie stated, not to be outdone. The giggling and laughter lasted all the way back to camp and throughout dinner that night. By the next morning the story was clear through the camp. I had taken a picture of a moose in my underwear. The variations of the story came back to me many times that day, always accompanied by uproarious laughter. Doesn't matter, I saw my moose.

Chapter 70
Duck Ale Grille

The restaurant at Space Farms Zoo and Museum has a number of different meats on the menu. When Gramma ran the kitchen (from 1933 to 1989), homemade soups, venison, elk, home-cured ham and bacon were the best of her home cooking. Today for efficiency and to satisfy the modern visiting family's appetite, the restaurant has become fast food. Beef burgers, turkey sandwiches and hot dogs, chicken nuggets, fries, onion rings and buffalo burgers are ordered by visitors on a daily basis. Breakfast sandwiches are served before 11 a.m. just like every other fast food restaurant I know. Space Farms has a number of very smart (the dumb ones fly into carnivore enclosures) free range chickens, ducks, geese and swans. Free range means the fowl are not caged up but are able to roam the zoo grounds at will to the delight of the visitors both young and old. Corn is sold to feed the free-range fowl. The first visitors in the mornings are often surrounded by fifty to seventy-five assorted birds looking for the special treat of whole corn. We feed the loose fowl a feed mix in the Barnyard Nursery part of the zoo. That makes all the fowl use the barn in the Barnyard to nest and roost at night. Chickens sleep so soundly that if you need to catch one, you'd best do it at night. You can sneak right up on one and pick it up. Ducks are a little more wary. Geese and swans stay on the lake next to the Barnyard Nursery.

Chickens come in hundreds of varieties. We have chickens that have multiple colors, sizes, feathery legs, fuzzy feathers, head bonnets of feathers, waddles and, of course, the roosters all have combs. Guinea fowl have polka dotted, black and white feathers and are known to eat ticks and bugs. Their clacky noises are annoying but, because they eat the deer ticks, we put up with them. Araucana American chickens lay colored eggs, pale green, blue and pink like Easter eggs. Rhode Island red chickens lay large brown eggs and are known for their egg size. Leghorns, like the famous cartoon character, lay large white eggs. Japanese Silkies are fluffy and so absolute white that they reflect moonlight and literally glow in the dark. The Silkies feather structure is such that they cannot fly. They are a favorite of the hawks and owls of Beemerville and of mine. The farms free-range policy on fowl means that our chicks are all crossbred and therefore are nature's most colorful mixture. Every few years we order from a catalog (yes, there is a catalog for farm fowl) to replenish the varieties that have flown into animal enclosures or the ones hawks and owls have carried off.

    Since the home base of the chickens is in the Barnyard Nursery, it is my job to collect the many colored and sized eggs laid every morning. The females are called hens. The males are called roosters or cocks. As a group they are called chickens. Chickens roam in a flock. A mature hen will lay an egg every morning during her cycle if she is fed good food that contains enough calcium and protein. The eggshell is made of calcium. If the eggs are left in the nest, the hen will lay eggs for a number of days, until she feels the nest is full. During this time the hen does not sit on the eggs. For some breeds the nest is full at a dozen, for some breeds two eggs are enough. Once the nest is full to the hen's satisfaction, a good hen will start to "set." Setting is the incubation period for the egg (21 days for chickens). During setting the hen leaves the nest only to eat and drink. This process seems to work out efficiently for the chickens and other domesticated fowl. If the hen started to set when the first egg was laid, the eggs would hatch out at different times, with the last chicks being left in the shell and the younger ones at a definite disadvantage. This does happen with many birds of prey, severely limiting the number of hatchlings that survive. If the eggs are taken from the nest, the hen will continuously lay one more egg every morning. Different fowl have different incubation times, ducks and geese-28 days, swans-32 days and ostrich-42 days.

    We have roosters, so all of our farm eggs are fertilized. I wish I had a nickel for all the times visitors run in and inform us that the chickens are fighting. Sometimes it is actually a cockfight; more often it is the hens and the roosters breeding. I try to explain this to the "informant" without embarrassing them, but the adult faces always turn red. The kids just think the birds are playing piggyback. If you crack open a fertilized egg you will see the fertilization spot on the top of the yolk, a small, white, translucent dot. The eggs of a free-range chicken have a richer colored yolk, bright orange, compared to the pale yellow yolk of the city store-bought eggs. The vibrant color difference is due to the higher level of protein, from eating bugs and worms. This fact totally grossed out my daughter when my dad first explained it to her, and she wouldn't eat farm eggs for a few years. The free-range farm

THE ZOOKEEPER'S DAUGHTER

eggs are richer (stronger) in taste and smell also.

The eggs we use in the restaurant at the zoo are the farm's own eggs. We rarely run out. Matter of fact we often sell our surplus to the city folk who visit the zoo. Many people prefer farm eggs. I put eggs in the cooler every morning, and the teenage kids in the restaurant know I get them from outside instead of from the grocery store like most of their parents. One day we ran out of eggs in the cooler. I don't remember what I was doing at the time, but all of a sudden I was called to the restaurant over the public address system. By the time I got in the building and was approaching the kitchen, there was quite a commotion going on. Seems two of the teenage boys had decided to take it upon themselves to collect the eggs needed for the breakfast sandwiches a customer had ordered. The boys had walked out to the nearest duck nest under a tree, shooed the duck away and brought in two eggs to cook. The hen duck was setting. The eggs were almost done with the incubation period. The boys brought in the two eggs and proceeded to crack one open over the grill. Flopping on the grill was one partially formed live duckling. The kids in the restaurant were totally grossed out, girls screaming and the boys appalled at what they had done. Jo Ann, the cook, had been on lunch break when the boys were "helping out." She reached the grill before me and scraped off the, by then, fried dead duck. I walked it outside and recycled the dead ducking by throwing it to the foxes, who quickly enjoyed the delicacy.

Dad had rushed into the kitchen also, and was raising holy Cain with the two boys. I calmed Dad down and gave the boys an "I'll-talk-to-you-later, mean mommy look." Later that afternoon when the lunch rush was over, I gave a "birds and the bees" (and the chickens, ducks and geese) talk to two shamefaced teenage boys who thought they knew all there was to know about sex. The other kids in the restaurant listened in attentively. No use yelling at them, the boys just didn't know. But now they do. I don't think their mothers minded that I gave the kids a talk on bird sex—I never heard a peep.

Chapter 71
My First Buck

I had not hunted deer since I was thirteen. Even then, I had never bagged a buck or a doe. Still, I, like millions of others, enjoy the taste of venison. I was raised on it, and it is good for you. Venison is not marbled fat meat. The fat on a deer is located on the outside of the meat, underneath the skin. It is, therefore, a low cholesterol red meat. When my family hunts and brings home the carcass, my dad is the family butcher. Dad's years of experience hunting and dressing out hunted deer has made him the best butcher I know. Plus Dad has gained experience dressing out road kills for the zoo. We always have a good supply of New Jersey's best homegrown venison with three hunters in the family. It also helps that we have a number of freezers amongst us. For obvious reasons we keep the road kill in the freezer in the feed house, not to be mixed in with the hunted-for-the-table deer. Just a note: After Doug told his mom about the road kill aspect of our jobs, my mother-in-law would not eat red meat at our table. She lives in Florida so we don't have to deal with what to serve often.

I had never bagged a buck. Oh well, no big deal to me. I couldn't care less; I didn't need to, since the meat was always there. I am involved with a number of local organizations. One of these is the New Jersey State Fair®/ The Sussex County Farm and Horse Show. Meetings are usually held at night so it was easy for me to attend after the zoo was closed for the day. The fairgrounds are located just down the road about five miles from Space Farms. The ride is a pleasant one on curvy country roads meandering through the farmland still prevalent in Sussex County.

I was on my way home from a meeting, enjoying the ride and singing at the top of my lungs to music. Come on, you know I have a teenager and my music time is limited—you probably sing in the car, too. (Go ahead, admit it, you'll feel better.) I know to watch out for deer. Space Farms picked up 1,984 dead deer from the road in the year 2000. We have trained ourselves, in this neck of the woods, to scan the sides of the road for the glaring reflective eyes of deer. I saw the first deer cross the road at dusk. The buck was behind her. I slowed the car to a stop, knowing that he would most likely follow. My red Jimmy was stopped in the road on the blind side of a curve. I checked my rear view and there was no one behind me, country roads are like that. The deer looked at the Jimmy, and looked across the road to his female friend. I proceeded slowly, thinking the buck was staying in the brush. I was wrong. I had barely hit fifteen miles an hour when the buck bounded in one jump from the left side of the road to the center double yellow lines. He skidded in front of my wheel with the dreadful thunk of doom. I hit my brakes again. My trusty Jimmy stopped dead in the deer's tracks. The deer was under the front of my car, thumpity thumpity, then miraculously dashed out and into the field. I pulled the car to the side of the road, rooted in the glove compartment for my trusty flashlight and shined the beam of light toward the bounding buck's direction. There, lying in the shorn hay field was the buck, thrashing in the throws of death. I cursed big time.

Nothing to do now but go home and tell Doug that a deer damaged the car and call Dad and tell him where to pick up the carcass in the morning. I was crying with the sudden rush of adrenaline by the time I got home. I rushed into the house, worried that Doug would be angry about the car. I should have known better. Doug hugged me and made sure I was all right, then we went out to inspect the minor damage done to the car. Not a big deal, the grill and headlight were damaged. We've seen worse done by deer and car interactions. Next I called Dad. Dad said he would be ready to go by the time I got to his house a block away from mine.

"Ready for what?" I asked.

"Well, he'll be good eating, he's a fresh kill," was his matter of fact reply.

"Allllll rrrright," I stuttered, "I'll be right there."

Dad and I drove the death mobile to the sight of the crash.

"Yep, that's a nice buck. But the hindquarter is busted up bad. The animals will enjoy the tenderized meat." Dad was almost jolly, trying to cheer me up. We loaded up the deer, brought it back to the feed house, dressed it out (took out the guts) and put it in the cooler.

The next morning, after I had my morning chores done, Dad asked me to report to the feed house. Standing there with a grin on his face was Bruce one of our zookeepers. He took my picture with the six pointer (if you count the tiny little nubs usually called brow tines). Dad bragged all over our little hamlet to all of his hunting buddies that so far I had bagged the best buck of the season. The good 'ole boys laughed on that one. After Dad reminded them about the turkey and the broom, the mice I've killed with a hammer (I was killing for a sick owl), the deer and the Jimmy, I was dubbed, jokingly an "alternative weapons master." Ribbing is always good-natured after the fact. The meat, well, the meat tasted just fine. I took home the loins, the privilege of the kill. The rest went into the family's larder. The picture sits on my bookcase, Best Buck of the Season. Yes it was.

Chapter 72
Tara Tiger's Tail Tale

On the Saturday before Thanksgiving, Karen Talasco, my friend and a great animal photographer, accompanied Doug on his grain feeding rounds. Karen is particularly fond of our pair of tigers and they have grown to know her through her photography. Any trip through the zoo Karen makes must include a visit to Khyber and Tara Tiger. I was home cooking lunch for my little family when Karen burst through our mud room door. She was obviously upset.

"Tara has a big rip in her tail; you'd better come look."

"How big?" I questioned as I grabbed my farm coat.

"Right through, you can almost see the bone," was Karen's worried reply.

We hopped into my car and were at the zoo in seconds. Sure enough Tara had a gash in her tail about a foot from her body. A golf ball sized section on the upper side only connected her softball diameter tail. Luckily the tail bones, vein and artery continue through the topside of the tail. There was no dripping blood so the vein and the artery were intact. I called to Tara to get her to come closer to me for a better view. She had been keeping the wound clean by licking, like most cats will do.

Karen and I hopped back in the car in search of Dad. We found Dad skinning a road-kill in the feed house. I told Dad I was going to call the vet. Dad agreed and said I'd better call quickly since Dr. Spinks was leaving on a hunting trip that weekend sometime. Dr. Spinks informed me he would be here within two hours. I had gathered the blankets we would need to keep Tara warm while she was under anesthesia. Doug had cleaned the tiger den and put down fresh hay. Khyber the tiger was locked out on the grassy section of the compound. Dad, Doug and I kept an anxious vigil while we waited for Dr. Spinks. We met him at the zoo gate. He drove his vet truck into the zoo and informed us that two of his assistants were meeting him there. Doug watched for them as Dad and I went out to the tiger compound with Dr. Spinks. Karen stayed and took Dr. Spinks' son for a walk in the zoo.

Dr. Spinks, Dad and I discussed the tail itself. The tail was still being held correctly, up turned at the end so the main nerve had not been affected. The lack of blood and Tara's seemingly lack of discomfort were in her favor. A tiger without a tail is a pitiful sight. Dad and Dr. Spinks talked of amputation for a second. Dad looked at my forlorn face. I was too involved emotionally with the patient to join the discussion. *Don't cut off my beautiful baby's tail.* The decision was made to try to save the tail. We could always amputate later if the tail did not heal. *Whew, what a relief!*

While still waiting for the assistants, Dr. Spinks prepared the anesthesia in a punch pole. Dad estimated her weight at 300 pounds. *My beautiful big baby girl.* I called Tara over to the side of the den. Dr. Spinks' aim was true. He injected her perfectly, right in the flank. The shot took effect in fifteen minutes. My job was to keep talking to Tara through her den bars to keep her calm and near the door. Finally Tara collapsed in drugged sleep on to a fluffy pile of fresh hay. Doug had escorted the vet assistants into the zoo while I was busy talking to Tara. Dr. Spinks was set up and ready to go to work.

Dad unlocked the tiger den door and walked in, picking Tara's massive, dinner-plate-sized hind paw up and spinning her around so her tail was to the door. Tara was heavily sedated, but the anesthesia would not last long. Speed and efficiency was necessary. One of the assistants held the end of the ripped tail gingerly on Dr. Spinks' orders. Deftly Dr. Spinks cut away the dead tissue while I watched Tara's eyes for response. Dr. Spinks stitched the tail together with heavy nylon sutures, which looked like fishing line. The assistants gave antibiotics and updated Tara on her other shots. Next the tail was wrapped in gauze and, of course, duct tape, the only thing that sticks to tiger and lion fur. I draped a blanket over the huge sleeping kitty to preserve her body heat. Pneumonia can be a problem in an anesthetized large animal after the anesthetic wears off. The whole process took an hour. We waited.

Dr. Spinks told us that Tara was a lucky cat. Another quarter of an inch and the vein or artery would have been slashed; Tara would have bled to death. After Dr. Spinks told me that, I really fell apart, and went for a walk to find Karen and Mathew. We returned to find the vet packing up and the assistants tidying up the area. Dr. Spinks stayed till Tara lifted her head, just to make sure she was coming out of the anesthesia. Tara was starting to come around. We all decided she would be better off left alone to rest without the additional stress of us standing there staring at her.

Dad and Doug walked the enclosure to see if they could determine what had happened to Tara's tail. We had tigers in that compound for thirty years and never had such an incident. They looked for blood on wire, or tufts of tail hair and could find nothing. Khyber was crouched in the far corner of the grassy section as far away as he could get from the lions next door. He was very wary. We attributed that to the activity with Tara. There was nothing to do but wait. Unbeknownst to me Dad was checking on Tara every hour. So was I. We bumped into each other on about the third hour. Tara was standing up and out of her anesthesia. She responded to my call and came over to the side of the den for pets and loves through the den bars. All was as well as could be expected. We went home for dinner. At dusk that evening I checked on Tara for the last time that day. She had successfully removed the entire duct tape and gauze bandage. But the stitches were still holding.

First thing the next morning I checked on Tara again. Not only was the bandaging no where to be seen, but Tara had licked out all of her stitches. One lonely thread of nylon hung from the bottom side of her tail. I rushed back to the office and called Dr. Spinks at home. We discussed the tail. It was a clean wound, the necrotic tissue had been trimmed off, and it was cool enough that there were no flies to infect with maggots. It was possible that the tail would slowly granulate back in. But she would have a scar. *A scar? I have a C-section scar, that's no big deal, if she can keep her tail. Tigers need their tails for balance while jumping and (maybe) their pride.* I decided to wait it out and see what happened. After ten days on food laced with antibiotics, the den door was opened and Tara ran outside to be near Khyber. Khyber had not left his corner of the grassy section. Doug (Parker was on a hunting trip) even had to throw Khyber's food to the corner. Doug had thrown chunks of hay to the corner so Khyber could sleep comfortably. The two tigers stayed in the hay bed in the corner, looking at the lions in apprehension. Khyber and Tara would not go into the den, as they had to pass close to the lions' entrance to their den. The young tigers were scared even though sturdy chain link separated the sections. Dad and Doug attached a temporary shed to the outside of the tigers' enclosure as the cold weather and snow was due to set in. The tigers immediately retreated into their new temporary den.

Doug and I discussed the strange behavior with Dad and between all of us we figured out that the lions next door must have been involved with the tail incident. Parker and Doug put a four-feet-by-ten-feet solid piece of tin between the chain link so the tigers and lions could not see each other. Two weeks later the tigers warily went back into their home den. I had raised chicken, wuss tigers. Tigers in the wild could take on the lions, no problem. The tiger is the absolute top of the food chain and takes no grief from anything.

My huge babies were scared of the neighbors. Miraculously Tara's tail healed completely. She doesn't even have a scar. The only clue that there was ever an injury on the bottom side of her tail is the three-inch long piece of nylon suture thread that still hangs from the sight of her injury. You can't see it unless you really look hard. Nature is a wonderment.

Chapter 73
Christmas Letter 2000

'Twas the night before Christmas and all through the zoo,
all the creatures are sleeping, so I'm writing you.
Hello to friends both near and far,
This year has been quite bizarre!
We survived bug Y2K,
Waking up happy on New Year's Day.
However, in March our computer died,
Day before deadline, I cried.
The 4-H Bear Safety book was due,
Doug and I worked fast I'm telling you!
Zoo opened early, way before May,
Then a blizzard blew in the very next day.
'Twas a wet summer, a dry one for me,
With hardly a soul born for my nursery.
Lucky the fawn was delivered by C-section,
when a doe crossed the road at an intersection.
Lucky is a sweetie now fully grown,
Next year she'll have young of her own.
Llama cria came late, none till August or so,
not a lion cub born, it's depressing you know.
I filled in at admissions and often the kitchen,
(where I do most of my bitching).
I love the zoo, the people, admissions are fine,
but I hate the kitchen all the time.
The summer flew by as they usually do,
there's plenty of activity here at the zoo.
Gave speeches and demos, talked up a storm,
of things that happen home on the farm.
I revamped the Eskimo exhibit late in the fall,

LORI SPACE DAY

A Styrofoam igloo surprised 'em all.
Painted windows on the old barn, sounds like no big deal
But the windows were plywood, not glass and steal.
Black squares on white wood, no great artist am I,
But I climbed the ladder 30 feet in the sky!
Doug is a saint, I don't know how he does it,
puts up with family and my temper fits.
He steadies my life, provides my balance,
when I get carried away with the "dance."
Jackie's become a social butterfly,
Straight A's and boys, God, I may cry!
She's growing up fast, much to my dismay,
I've become a taxi, the mall is which way?
I hate the mall, it's just not for me,
But our time together is a precious commodity.
We vacationed in Canada, I mooned a moose.
Got a picture sans pants on my caboose!
I'm writing a book for those who don't know,
of my life's work with the animal show.
Stories, adventures, mishaps and tragedy,
with all of my animal friends and me.
The "Family" is nervous, I've not let them look,
"Exactly what is in THAT book?"
Got my first buck last day of the season,
suicide by car, what was his reason?
I'm an alternative weapons master–
Remember the turkey I killed with a broom?
The picture now decorates my living room.
The guys all laugh at my weapons choice,
Doug didn't even raise his voice.
So here is my annual Holiday poem,
call, write or visit, I'm most always home.
Love, Doug, Jack and Lori

Chapter 74
Bottle Breaking Multiple Species

With spring just around the corner, the nursery at Space Farms Zoo will be occupied with animal infants that need human intervention. My dad, brother and husband handle the larger animals. The Space Farms Zoo has over one hundred different species. The nursery is my responsibility, my calling and my joy. If at all possible we leave the young with the parents, mother's milk and care being the best that nature can supply. Sometimes infants need human help due to inadequate maternal behavior or medical problems. The infants may be designated to another zoo for genetic diversity. Whatever the situation, the infants come to "my" nursery. If you are thinking of hand rearing any animal specie, be prepared for the commitment of time and care it will take. If you are not willing to make that total commitment, the baby, no matter how sick, is better off with its mama. Nature will take its course.

The first step is removing the infant from the mother's care. We all know this is easier said than done with some species. With difficult species, we at Space Farms Zoo prefer to use the "bait to separate" method. Bait can be anything—food, disturbance of the outside enclosure (the "maternal protection of the nest" instinct challenges the mother to leave the young), or the cry of another infant of the same species. Bait can be anything that will bring mama running to the rescue and away from the nest. Hoof stock herding species require multiple manpower and/or vehicles. We prefer not to use chemical immobilization due to the absorption of the chemical into the milk for the siblings that may not be removed, or, the infants to be removed may nurse before mama is completely under. A drugged mama may fall on the infants and crush them. Be careful, have a plan, and remember no animal infant is worth dying for. Mama is not going to appreciate the help you are planning to give her baby.

After you have the infant in your care a quick assessment of the physical condition will tell you if this is a case to take to the vet. If you have any doubts, a vet check will put your mind at ease and medication will help a sick infant.

Bedding and warmth are next. I prefer towels for the species born with eyes closed. Zooanotic diseases require separation of different species or sickly siblings. Yes, towels require washing, but they are soft, warm and fuzzy, like mama. Warmth depends on the specie and age of the infant. If the runt of a litter is constantly being laid on or sucked on by its siblings, rear it alone or with a small littermate.

We place infant cubs and pups in clean cardboard boxes (easy to come by and disposable afterward). If you cut another piece of cardboard the size of the bottom and cover it in a taut heavy plastic, it will act as a great liner so urine does not soak through the toweling and into the box itself. The liner is then placed in the cardboard box. A heating pad is placed in one corner and the liner and heating pad is covered with a towel. Cut a hole in the box to thread the electrical cord through so the infant does not become entangled. Leave the top on the box to close it. This keeps heat and the critter in, and provides a dark, cozy space. Make sure your box is big enough to allow the infants to crawl off the heating pad if they get too warm.

Small hoof stock is kept in the same set up. Skip the toweling and substitute ground corncob for absorbtion, hay for chewing and traction for standing. Inside room temperature is usually sufficient for these young. If your infant is sickly, add heat. For these I prefer a heating pad vertically taped to the side of the box (to simulate lying next to a warm mom). Dad and I have seen animals that we thought were "goners" come back within an hour under a heat lamp. Heat lamps are tricky, suspended too close over hay is asking for trouble. The standing infant's highest point should be eighteen inches away from the heat lamp, to prevent burns.

Hopefully, you are prepared for the type of infant coming to you. Everyone has their own formula preference, whether a home-mixed formula or an established commercially prepared product. This varies of course from specie to specie and time saving necessity. We like Land of Lakes Kid Goat milk replacer for hoof stock (llama, fawns, goats), sheep milk replacer for sheep, and Blue Maxi for calves. We use the Zoologic brand powdered milk replacers for carnivores and omnivores. We are fortunate to be located in a dairy farming area. The local farmers supply us with colostrums, the first antibody rich milk of a cow, which we start our hoof stock on if they have not nursed from their mom. Eight ounces for fawns, goats and sheep, then we wean them, mixing with the prepared milk formula. I have found the Borden/Pet Ag Company very helpful if I had any questions on formula for species we've not hand-raised before. They will send you information on the percentages of solids, fats, proteins, and carbohydrates in comparable species.

All of this advice does not matter if you cannot get the formula into the infant.

After the psychological shake up of being removed from mama, being given a physical, then trundled around and placed in a warm, cozy but strange box, I give my new charges a couple of hours of peace and quiet if possible. Time, quiet, dark, and warm will settle a youngster enough so he realizes he is hungry. Newborns are the easiest to bottle break. The natural instinct to suck is in our favor. I prefer a cross cut human baby nipple. It will hold milk in with less mess when upside down and gives free flowing milk at the gentlest suck. Start with a small cross cut and increase the size of the cross, as the infant needs. Common sense applies here since you do not want to drown the baby or cause him to aspirate the formula. The best nipples are those with a bulbous base, especially for animals with a longer snout. If the nipple constantly collapses flat, use a stiffer nipple, sold as older child nipples in the stores. Calves and larger do well on a calf bottle available at most animal supply stores.

With litters I color-code the rim of the bottle to the collar on the animal. This helps to make sure how much formula each infant is receiving. A thorough hand washing between individual babies prevents passing along germs from one to another.

For cubs and pups, I hold the bottle by the rim between the thumb and first finger, nipple pointing INTO my palm. I place the pad of my thumb gently over the nose, to simulate the warmth of the mama's breast. Using my palm to cuddle the wobbly chin, and holding the head in an upright position simulates resting the chin on the breast of the mama while nursing. In the beginning I try to get the infant to suck on a finger and slip in the nipple. I keep my ring finger fingernail clipped short if I need to pry open the baby's jaw (gently!) to get the nipple in. After a few feedings the infant usually catches on and will open its mouth at the first nudge of the nipple. It is our rule of thumb to feed an infant from the start, consistently, in a position that will enable you to feed the youngster as it grows. Feet down, head up on a non-slip surface, lines up the baby's body like it was nursing from mama. Nothing is worse than a big fifty pound plus baby that thinks he can drink a bottle sitting on your lap. Feet down from the start will help train that baby, avoiding problems later on.

Newborn hoof stock will usually nurse right after being cleaned off by mama or you. I use the same position again, holding the rim of the bottle between the thumb and forefinger, resting the chin on my palm. It is very important with hoof stock to keep all four feet on the floor. A squirming baby may fall off your lap causing injury to itself, or get its formula down the wrong pipe. Aspiration of formula can cause pneumonia.

At this point you may wonder what I'm doing with my other hand. The hand not holding the bottle is in constant motion, rubbing and massaging the body of the infant. Again, this simulates what the mama would be doing, licking and nudging her newborn, encouraging him to nurse. After the feedings, we burp (hoof stock burp themselves) to avoid gas pains and spit ups. We also stimulate for urine and bowel movements, using a wet paper towel or wash cloth. Mama would lick her baby, but there is only so much I will do for my job!

Newborns will drink till they burst if you are not careful. We feed newborn African lion cubs 2 tablespoons every two hours to start. They grow fast and by a week's end are usually up to two ounces, skipping the 2 and 4 a.m. feedings. By four weeks lions are up to five ounces and given raw ground meat (venison). At six weeks our cubs get eight ounces, five times a day, plus raw ground meat. Many people in the trade feed cooked meat and are surprised that we feed raw meat. Again, we imitate nature. Mother Nature does not cook meat for her wildlife. Cubs and pups in the wild would lick, suck and chew on what mama brings home.

Fawns get two ounces to start and are gradually increased to eight ounces spaced out six times over the day. Llamas will start at six ounces and increased to eight ounces six times a day. Other species are handled proportionately. It has been my family's general rule that it is better for a baby to be a little hungry and fed more often than to take the chance the infant will engorge and suffer.

Newborn animal infants are the same as human infants: they eat, poop and sleep. I prefer not to over-handle a newborn; they need their energy to grow. Gentle stroking and rubbing during feeding imitates mama's actions. There is plenty of time for cuddling and playing once they are more mobile. You will know when they are ready to play. Let them play first. Sometimes love means doing what is best for the infant, not for you. You must love them or you wouldn't be doing this time consuming job!

Older youngsters that have been nursing from mama are a bit more challenging. It often takes twelve to twenty-four hours for an infant to get hungry enough to overcome its fear of you and the new surroundings. Healthy infants will be just fine during this time, though they may whimper till they get hungry enough to try the bottle. I give them a try at intervals of four hours. Sickly infants may need veterinary help.

I had problems with three-week-old coyote pups once and accidentally discovered a trick that may help you. They were hungry, but would not suck. Faces got messy covered in milk and I handed a pup to my daughter to clean off its face. A warm wash cloth was used and the pup latched on to the cloth and would not let go. My daughter brought this to my attention and I realized the pup was sucking. A pair of scissors and a hole in the middle of a warm wet washcloth with a nipple sticking through it solved the dilemma. The pups nursed with the wet washcloth over their heads. They may have looked goofy, but they drank!

Peace and quiet can help in the bottle breaking process. A youngster is easily distracted from the learning job at hand by strange sights and sounds. Selecting a quite area to feed can eliminate the sounds. After the youngster is adjusted to the bottle, other sound distractions will not bother him. Leave your cell phone someplace else while feeding. The sudden ringing or vibrating of a cell phone can upset an animal learning to nurse from a bottle.

Covering the eyes of a youngster also eliminates distractions and/or fears. I find this especially helpful with hoof stock that have nursed from mama. We keep fawns quiet and secluded for a day or so until they are on the bottle. Fawns are the only exception to the four-feet-on-the-floor rule. Nature tells them to lay still and hide. I use this and quietly approach them, letting them stay in the laying down position. Using one hand to cover their eyes (their eyes are closed when nursing between mama's legs), I pry the nipple into their mouths and stay quiet and wait. Healthy fawns will suck, recognize the bottle, and stand quietly nursing after a day or so. This applies to white tail fawns; sika and fallow can be extremely difficult if they have been with mama.

Llama cria can be challenging. Newborns are the easiest. We leave the cria with mama for one week to ten days if they are healthy and scheduled to go someplace else. When I get them they are scared. I leave them alone for six hours to get used to heir new environment; even chickens can be scary if you've not seen them before. Llamas cush (give up and collapse). I use this natural phenomenon while bottle breaking llamas. I slowly approach the baby and wait till the last possible moment to reach out and grab a hold of him. I use both hands to settle him down and swing one leg over his back. I seem to be just the right height so I am not riding them with my body weight, but corralling the cria between my legs. With one hand rubbing the neck, the other hand slowly comes up with the bottle between the thumb and forefinger, nipple toward the palm. Using the middle or ring finger of the bottle hand I gently pry open the jaw behind the front teeth and insert the nipple. If the cria does not suck right away, I rub the neck area. Leaning over the baby and shading/covering his eyes with your body helps to eliminate distractions. I often look like a contortionist while starting these babies on the bottle.

Weaning age depends upon the species and the bonding time you have available. We often continue a token bottle as a treat.

These methods have worked for us for three generations of Space family. My dad and grandparents taught me more about animal care, hands on, than I learned in college from a book. We are not perfect and occasionally lose young like anyone else. That can be heart breaking, but not all babies are destined to survive, even with their original mothers.

Teaching young Dolly llama to drink from a bottle was a challenge.
Photo credit: Karen Talasco

Chapter 75
Rescuing Jane Doe

On Sunday May 27th, Space Farms received an unusual deer call from Donald Howey, a farmer a few miles away from the zoo. This deer was not dead on the road, but very much alive (and upset) in the bottom of Mr. Howey's well pit. Mr. Howey had discovered the deer in the eight by eight concrete underground bunker when he had investigated why the water was not working on his farm. Mr. Howey turned off the electricity and called the Spaces. Dad and Parker went to check out the deer's situation. Parker visually inspected "Jane" Doe for broken legs. She apparently had none. Mr. Howey, Dad and Parker conferred and the decision was made to tranquilize the doe in order to hoist the desperate animal out of the ten-feet-deep well pit.

Parker called home to the zoo giving the weight, sex, and size of the doe to me. I relayed the information to Dr. Ted Spinks of the Animal Hospital of Sussex County. Dr. Spinks prepared and donated the dosage of tranquilizer for the deer.

Parker, who was the current chief of the Beemerville Fire Department, called upon his firefighting team for assistance in the rescue. John Stearns brought rescue straps to wrap around the distressed doe. Dominick Gallo, his son Vinnie (with a video camera), Scott Haggerty, Irish Dunn and Mike Turo completed the fire and, now, rescue team. Members of Mr. Howey's family were also on hand to help. Dad asked me to go along in case the doe died and a c-section was needed and to record what happened to send it to the papers as a press release.

Parker prepared a punch pole (a long pole with a hypodermic needle on one end), while Dad prepared the tranquilizer gun, under the supervision of Dr. Spinks. Fire Chief Parker instructed the rescue team to wait at a distance as he descended the rickety ladder through a three-feet-by-three-feet shaft into the well pit. The deer was frantic and dangerous in her panicked state. I saw the panicked deer thrashing around in the small underground bunker and let me tell you it took a brave man to enter the small opening to the well pit. I have carried yearling deer that were somewhat tame and did not want to go to where we wanted them to go. The tame yearlings can kick the bejesus out of you. Parker showed true courage entering the eight feet square concrete bunker with a wild adult doe that had no way out and no understanding that he was there to help her. My respect for my "little" brother increased that day. Cautiously, Parker waited until the doe was in position and jabbed her with the tranquilizer in the hindquarter. A sigh of relief was heard when Parker climbed out of the well and said, "I got her, now we wait fifteen minutes for the tranquilizer to take effect." Everyone readied for action. After leaving the frightened doe alone for the fifteen minutes, the Spaces decided it was time to haul the doe out of the pit.

Parker once again descended the shaky ladder and discovered that in her thrashing around, she had wrapped the electrical wires from the eighty-gallon water tank and pump around her neck. The wire had to be cut before she choked to death. Again Parker took on the dangerous job, hanging off of the ladder with a pair of wire cutters. The dosage of tranquillizer was not enough to put the doe completely under (too much tranquilizer will kill), just enough to slow her down, enabling Parker to get the rescue straps around her. The deer was still frantic, pacing in the slippery, tight quarters. Parker was able to get the lassoed rescue straps around the doe's head and fore feet. A gentle tug snugged up the rescue straps around her abdomen.

On command the rescue team hoisted the deer up and out of the ten-feet-deep cement prison. As Jane Doe immediately bounded to her feet, the men quickly loosened the rescue straps. Jane Doe groggily ran off into the woods, with only a few scrapes, bruises and *I am sure* a bad hangover the next day.

Upon inspection of the well pit, the doe had miscarried two fawns due to her fall. The fawns were premature and dead. Dad thought that might be the case. Mr. Howey made plans to cover the well pit more securely to prevent another, possibly more injurious, accident for man or beast.

Cheers and congratulations were given to all the volunteer firemen, who gave up their Sunday afternoon of a holiday weekend to help out Mother Nature. And I discovered my brother, my favorite little pet in 1969, had grown up into a very brave, strong man.

Chapter 76
Three Inquisitive Raccoons

The raccoon is a fascinating animal. It is considered a pest in the local suburban community, constantly raiding garbage cans and taking up residence in attics in human seasonal lake homes. This is a testament to the unique ability of the raccoon to adapt to the encroachment of its territory by humans. In the wild the raccoon is an omnivore, eating whatever it can find from fish, frogs, mussels and crayfish to berries, grubs, nuts, and farmers' corn. Suburban raccoons love people garbage. It is easy pickins', less work, more variety, and higher sugar content. Raccoons, like humans, love their junk food.

Wild raccoons establish a routine of travel through a territory, visiting certain spots where they know food is available, often by streams or bodies of water. They may travel miles on the circuit of their territory. Males do not establish a nest, sleeping for the day in the crook of a tree wherever they end up. Females travel less far from an established nest den in a hollow tree or abandoned squirrel nest. The availability of food is the determining factor. Raccoons are nocturnal, traveling from dusk to dawn unless they are sick. So if you ever see a raccoon in the daytime, steer clear of it.

Suburban male and female raccoons often use the same den or sleeping spot day after day. Food is easier to find in the garbage can; the pet food left out for household pets, or a compost pile. Raccoons are great climbers, the claws on the feet grabbing into trees or the sides of houses or chimneys. They can bound up a tree in seconds, using both front feet at the same time if in a hurry, or one foot at a time if they are not frightened. Like squirrels, the raccoons travel up and down a tree headfirst, but can also back down a tree if so inclined.

Known for their distinctive markings of a bushy striped tail and dark mask on their faces, raccoons are the night raiders of suburban society. The fur of the raccoon is actually three colors on each guard hair shaft, dark brown to black gradually coming to a silver tip. The under fur is usually dark brown. Many color phases exist, however, some much lighter than others.

The darker raccoons occur in the northern parts of their range. The raccoon ranges throughout the United States and the southern parts of Canada. Raccoon fur is soft silky and used for fur coats, collars and muffs. Trapping and hunting of raccoons is legal in most states.

Suburban raccoons cause a lot of damage to the houses and buildings they inhabit. A raccoon is a good-sized animal, the body about the size of a basketball on a twenty-five-pound animal. Raccoons have been known to reach weights of up to fifty pounds. The destruction of property and the night noises of raccoon life activity cause the suburbanite to call the specialists in. That is my Brother Eric's business, he is a suburban trapper. He live traps and removes the animal to a wilderness area. All raccoons do not have rabies, but that is another common fear of the uniformed suburbanite. Wild raccoons are nasty animals, with long, sharp claws and wicked teeth, not to be handled by non-professionals.

In the spring, raccoon babies are born after a sixty-five-day gestation. The young are hairless, the skin having a slight coloration in the pattern of an adult. By one week old the fuzzy fur in a soft brown color is coming in. The infant eyes are sealed closed, but open at about three weeks old. By this time the baby raccoons are quite agile, climbing all over the place. The litters average four to six babies. The more babies the scrawnier they are. I've seen some pitiful-looking raccoon babies.

Raccoon pups make a number of noises, constantly. They hum, chipper, grunt and growl, especially when they are hungry. When Eric live traps a female raccoon out of a house in the spring, he checks if she is nursing. If possible he releases the young with the mother in the woods. Mom then takes her babies to a new nest site. In the wild, a mother raccoon will move her nest if it becomes soiled. This process takes time as mother raccoon can only take one baby at a time, carried by mouth. Often hikers come across young raccoons that have wandered from the nest while mom is transporting the littermates. These babies should be left alone; mom will be back for the rest of her young. Mother raccoons are very maternal and protective of their babies, rarely leaving them longer than it takes to find food.

Years ago it was legal to have raccoons as pets in the state of New Jersey. Raccoons do make loving pets for the first year. After the coon reaches sexual maturity, most raccoons "turn" and get nasty. It is no longer legal to have raccoons as pets in New Jersey. It is a good law protecting the unknowing public from the possibility of rabies and the "turning" process.

My brother Eric had a routine trapping assignment from a suburban attic.

The raccoon he caught out in his live trap was a female, but did not appear to be nursing. He released it in a wilderness area of Space Farms. A few days later the inhabitants of the home heard unusual noises coming from the rafters of the attic and called Eric back. Eric found three little raccoons, maybe three weeks old. Eric brought them to me, would I help them out? The mom had been released earlier and was nowhere to be found. What could I say?

I kept the three adorable babies in a cardboard box with a heating pad. I put them on Esbilac, the milk suggested for young raccoons. The raccoons loved their bottles and quickly bonded to me. After a ten-day quarantine, and very wary handling, Abby, Bobby, and Chuckie (A, B & C), two boys and a girl, were put in the zoo nursery. They grew quickly and by five weeks old were eating solid food, along with their daily bottles of special formula. At five weeks old the pups were about the size of a Nerf football, and about that round. They had grown in the colored fur of the adult raccoon and the trademark eye mask and striped tail. Cute and full of personality the three Musketeers climbed everything they could find. Their dexterous fingers sought out the shape of anything they touched. Abby, Bobby, and Chuckie would play for hours with the baby toys I put in their kiddy bathtub. Raccoons love water, but after playing in the water the adorable little raccoons looked like wet rats. What a difference in their appearance! Image is everything. Dad and I call it the fuzzy factor, the fuzzier the animal, the more cute and adorable it is considered. Snakes, spiders and lizards have no fuzzy factor. Lions, llamas and raccoons have a high fuzzy factor and are loved by the public. Regardless, the three babies were really cute.

Hunter and Lindsey Space and my daughter were the only children allowed to play with the young coons. The possibility of a visiting child getting bitten and suing us is always foremost in our minds. One day Hunter and Lindsey were feeding the young coons animal crackers through the door of their enclosure in the zoo nursery. Dad and I were at the admission counter right inside the door to the zoo. All of a sudden, the zoo door burst open, Hunter running in out of breath.

"Lee-lee, the coonies are out, the coonies are out!" Hunter stated, obviously upset. Dad and I calmly walked outside to go get the raccoons and put them back in the enclosure. Lindsey was by the enclosure next to the door, looking sheepish.

"Who opened the door?" I asked.

"I did, Lee-lee, I was feeding them," little Lindsey replied.

"Well, we better catch them," I said, as I picked up the first one from the green grass and put him back.

"I can get them, Lee-lee," I was surprised to hear five-year-old Lindsey say. And she did. Without hesitation or worrying about getting bitten by the squirmy climbing raccoons, Lindsey got on her knees under the bottom of the enclosure. She deftly grabbed the small raccoon in her smaller hands, the two handed baseball scoop style I had taught her. Gently but firmly Lindsey placed the rebellious raccoon back in his house.

"Stay," she commanded, even though I had not taught the raccoons that command. Then she climbed back under for the last raccoon.

Dad tapped me on the arm. "She has the 'touch,'" Dad whispered to me.

"I know, Dad," I confirmed. Working with animals all our lives we have discovered that the "touch" is something special, something you either have or don't. A genetic aura if you may, that lets you communicate with or "read" an animal. It is hard to explain as all special gifts are.

The little raccoons stayed with me throughout the summer. In early fall the mamma raccoon would chase the youngsters away to live on their own. That fall I let the three raccoons go in the stream in the forest at the other side of the Space Farms property, away from the zoo. The raccoons quickly forded the stream looking for crayfish with their underwater hands. They never looked back. I knew I had to let them go. They would squabble if I put them in with the zoo raccoons. I went back the next day to call them, just to see them and give them a treat, but they were nowhere to be seen. I know they are doing fine, there is a lot to eat out there and plenty of hollow trees for shelter.

Chapter 77
Fritzy,
the Black Australian Swan

Al Rix stopped by for coffee with me one day in June. He had just brought a load of bread in trade for a load of fresh venison for his bear ranch in New York State. Over coffee we gabbed for a few moments about his family and mine. Out of the blue Al said, "I have a svan, er name ist Fritzy. I vaised er vom a cheeck. See sleeps every night on mine chest. Yov vant a black svan?"

"Sure, Al," I responded, "I can put him on the pond."

"No dat vill not be good, yov keep er in the yard, yes ok? See is a girl, but er name is Fritzy" I knew he was referring to the Barnyard nursery.

"No problem, Al, I'll keep her in the barnyard. How big is she?"

"See is a year old, I raised her, see is my baby. I vill bring er zoo you."

Al told me the story of Fritzy. Al's black Australian swans on his home pond hatched Fritzy out. Raccoons came in and got the other sibling cygnets (baby swans). Al's aged eyes shown with love as he continued Fritzy's life story. Fritzy was brought in to prevent the raccoons from getting her also. Fritzy was pinioned (one wing is cut so they cannot fly), and fed on lettuce and swan pellets. Al tried to reintroduce her to her parents, but the parents were not friendly to Fritzy. The yearling swan was living in the house with Al at night, going outside to a playpen in the daytime. Al told me how every night the cygnet swan would sleep on his chest and follow him around every day. You could tell Al was very attached to his swan, I was honored that he offered it to me.

Al told me what I would need as far as food was concerned. I told Al that we had a kiddy pool I used for the otters. "Yez, I heard 'bout zour otter, I am zorry." Al referred to Lizzie as he patted my hand. It was a long time ago. I was surprised Al remembered. Al went on his way. My day was brightened by his visit. His jolly personality and upbeat attitude was always easy on the ears.

Black Swans are indigenous to Australia and New Zealand, and are black, hence the name. Both the cob (male) and the pen (female) have white feathers tipping the wings. The outstanding red beak becomes even more flamboyant during the breeding season. Capable of flying up to 50 miles per hour, they do not migrate and will stay on a home pond as long as food is available. With a wing span of seven feet, the adults weigh twelve to fourteen pounds and are a big bird. Swans mate for life, taking turns on the nest, while the spouse protects the area. This fact makes them very desirable for home ponds as they chase off the Canada geese that have become green pooping pests here in the Eastern United States. The Black Australian Swans eat aquatic vegetation, fish and corn. An average clutch of swan eggs is 5 to 7. The pair actively protects the nest from all intruders. These large birds can be very intimidating, clobbering you with wings and making a lot of threatening noises. The cygnets hatch after a thirty-two-day incubation period. The small, gray, fuzzy cygnets are the size of a grapefruit when first hatched. It is necessary to keep the young off the water until they are older as the cygnets do not produce the oil that will waterproof their feathers. This is why you see wild swans carrying their cygnets on their backs while swimming. Young swans can drown. Once the oil glands develop the cygnets swim after mom and dad. At that point the main killers of swan cygnets are snapping turtles. Yes, we have them in our zoo pond also, all part of the cycle of life. The Space Farms Zoo pond was protected from feral dogs and raccoons so we did not have to worry about those potential predators. .

A month later, I had forgotten about the baby swan, Al pulled into the office as I was walking by.

"I 'ave zour new baby!" he exclaimed excitedly. I went over to his truck and peered into the back. Al's "baby" was huge! Definitely not the volley ball size I expected. Guess I didn't know as much as I thought I did on swans. Fritzy was easily two feet long and three feet tall to the top of her head. Al and I set her up in the nursery with the otter kiddies' swimming pool. Fritzy jumped right in before we had all the water in the pool. She squeaked and squawked, honked and hooted. I did not know that black swans made so much noise. The white swans we had are mute. Her squeaky noises were nifty, like a dog toy. I stroked the juvenile swan like Al showed me. She hooted and called to us, stretching her long neck in pleasure.

"Goot, see is happy." Al patted me on the back. "Zou vill take goot care of er?"

"You know I will, Al, she is really neat." I couldn't help gushing.

"I vill bring zou a boyfriend, OK?" Al was misting up slightly.

"Yeah, sure," I replied. Al and I had coffee and then checked on Fritzy from the back glass door of the zoo. She was happily swimming in her much too small pool, honking at the surrounding chickens.

A week later Al was back with two more swans. It was all right with him if we put these two on the pond. So Al and I released the new black Australian swans on the acre pond bordering the nursery. Fritzy would get to know her new neighbors through the fence first.

Two weeks after the swans were settled in, Al died of a heart attack while tending to his beloved bears. I think he knew that it was going to happen, in some way, and wanted his Fritzy to have a good home and a future mate. Fritzy's parents are still on Al's home pond. Al's son and daughter are taking great care of them. I know they will help me if I need advice on black swans.

For the summer months the black pair swam the pond and wandered through the zoo, honking at visitors to feed them. Fritzy stayed in the nursery till October, and then I opened the fence gate between the nursery and the zoo pond. The adult black swans accepted Fritzy readily and the threesome was often seen frolicking the Swan Lake ballet across the pond.

After we closed for the season on October 31, the guard gates are left open so the men have easy access to the animal enclosures during the winter snow. Early November Dad met me in the morning with bad news. One of the black swans had gotten too close to the jaguars' enclosure and was nouveau cuisine. *Some meal, black swans are worth $450.00. Oh my God, Fritzy!*

I rushed out to the pond and called to Fritzy. I hooted and hollered, calling her by name. Finally after what seemed like forever, she called back and swam across the murky pond to me. I joyously tossed her the crusts of bread she loved so much. I was upset about the dead black swan, but at the same time euphoric that it was not Fritzy.

I reported back to Dad and we decided to set up a temporary pen by the pond next to the nursery to house the other black swan and Fritzy. By doing this we hoped to establish the swans in one spot where there was a shelter for the winter. Doug moved the food trough from the nursery to the pen by the side of the pond. If the swans learned where the food was, hopefully they would stay in that area. The swans were kept penned up there for a month as training. It seems to have worked. The two black swans stay in the section of the pond by the nursery. Fritzy greets me every morning when I call to her. Doug calls Fritzy my squeaky toy, but the sound of her voice is music to me every morning. I open the gate to the nursery and she trots right in like she owns the place. I'm not sure which of the other black swans was dinner for the jaguar, the male or the female. So I don't know if Fritzy has a mate or a girlfriend. I don't care. I am so happy to have Fritzy.

Nephew Hunter helps to feed young Fritzy,
the Black Australian Swan.
Photo credit: Karen Talasco

Chapter 78
Pushing the Buck

White tail buck deer fawns raised by human hands often grow into a problem for humans. The problem is psychological rather than physical. Raised from fawnhood by humans, a buck has no fear of humans. Male deer go into a hormonal season that is called a "rut" during September through December. This corresponds with the estrus season of the female. Amazing how nature works this all out.

Bucks grow antlers, not horns. The antler growing process starts in the spring. The antler itself is actually a calcium deposit. Velvet is a system of veins and arteries that feed the growing antler and small hairs that protect the antler while it is forming. The velvet begins to shed in late August, and the process must be itchy, as the bucks will rub their antlers on anything around. During this time the bucks look very unkempt with tattered velvet hanging from the antlers on their heads. There is an old wives' tale that says if a buck is injured on his body, the antler on that side of the head will grow crooked until the body is repaired. Dad, my brothers, Lenny Rue and myself have found this to be true, but I have never seen it in any books or scientific literature.

In autumn bucks become territorial, fighting each other with the antlers grown since the last spring. The buck's neck becomes "blown out," stronger to hold and fight with the pointed crowns of calcium. Throughout this section of New Jersey the woods are often filled with the clickety-clack of antler fights. It is a distinctive sound. The winning male gets to breed the local does. Bucks have had fights in the wild where the antlers become locked and entwined and the bucks cannot separate. These bucks die. Bucks are gored and injured during this fighting phase. When all of this happens in the wild it is Mother Nature's phenomenon. When the buck deer are in the zoo herd, it is our problem.

Dad tells the story of years ago how a buck in rut got him down in the deer run during the rut season. Dad was thinner then, *Sorry, Pop,* and was able to straddle the antlers around his waist, the deer pushing him toward the gate. Dad was able to escape the run unharmed. Many other zookeepers in zoos have not. When the animal folk get together with Dad at our zoo and start to gab, they are often like little kids comparing scars.

The New Jersey Division of Fish and Wildlife brought Space Farms a white tail buck that had been illegally hand raised by a civilian. The problem buck had been chasing humans. The Division confiscated the buck and brought him to us. While the buck was under anesthesia his antlers were sawed off. It was a good thing they did. Doug went in the acre run to feed the deer grain, like he does every day. This involves lugging three fifty-pound bags of ground corn and oats. Farm work has made Doug a tough, strong guy. The new buck came up to Doug and pushed Doug with his head. Doug shoved him away. The same buck came at Doug again, harder. Doug backhanded the buck. That was all it took to enrage the hormonally charged buck into fighting mode. Using his fist and forearm Doug whacked the buck away again as he hurried to the gate. After Doug was out of the fencing, the buck came at the fence, headfirst. The clanging of his head on the metal poles did not seem to phase the buck. Doug's forearm and fist were swollen black and blue for a week. Every day after that Doug took a baseball bat in the run with him. If the buck saw him, the buck would come charging. Doug learned how to feed the deer in record time that year.

By spring the antlers are shed naturally by the bucks. The antlers just fall off, just like a five-year-old human loses teeth. No pain, little blood and very little problem. The small nubs of the confiscated buck's sawed off antlers dropped off naturally. Out of the rut season the buck was no problem. Gentle and friendly he meandered the grassy acre. Come that fall, antlers grown in once again, the buck became a problem once more. Not afraid of humans, the buck was becoming aggressive toward the men in the beginning of the rut season. The antlers had replaced themselves to their natural glory, a nice eight pointer that would grace any hunter's living room wall. The confiscated buck was a real danger to humans now; his eight spears a weapon ready to impale the trespassing human or other bucks alike. Doug was aware, however, and avoided the buck, sometimes asking other zookeepers to "rattle" the dangerous buck away from the gate. Eventually, we had to shoot the buck, as he was charging the fencing. The problem buck was charging the human on the other side of it, the fence just got in the way. We could not take the chance of a person being gored with the tines of his antlers through the fence. The adverse training Doug had tried failed. You cannot fight Mother Nature.

This is a prime example of why the state of New Jersey has laws that prevent unknowledgeable "civilians" from raising white tail fawns.

Gramma and I had raised white tail deer fawns from our deer run since, well, forever. At the zoo, the friendliness factor is important. However, when we raise the fawns, Gramma's and my involvement with the deer is exclusive. We break off our relationship with the fawns in October. When the does in the wild stop nursing their babies, we stopped bottle feeding ours. By January the fawns, now half grown, are skittish once again, avoiding the humans that enter their run. This seems to work well; the young bucks we reared have never given the men any problem. I, however, do not go into any of the deer runs come September when the rut begins.

In mid-August, Carol Decker, a renowned artist and naturalist, wanted to shoot video of white tails and other native animals for a TV special. Carol Decker has a special place in her heart for the Space family and the zoo. Years ago when she first started painting, she sold her paintings in our gift shop. The Space family is proud to have been a part of Carol's painting history. Carol and I have the common interest of painting although I could not in my wildest dreams attempt to be as good an artist as she is.

Carol and her videographer came to the zoo and shot footage of the raccoons, Abby, Bobby and Chuckie with zoo visitors. Next we took video of that spring's fawns, and two native snakes all without incident.

It was mid August, so when Carol asked if we could shoot video in the white tail deer run, I figured it was early enough that the rut season had not started. It would be safe to do so. Most of the deer in the run are friendly, but skittish, staying away from us humans, with the exception of Lucky the imprinted female Caesarian fawn from the previous year. I was confident that Lucky would give the videographer the interaction he was looking for. The three of us entered the grassy acre run.

Lucky was the first to bound over to me, happy to hear my voice and munch on the saltine and animal crackers I had brought with us. Deer love salt. First we stood in the waist tall grass, and then Carol squatted on the ground. I gave Carol all the goodies and stood aside, hoping Lucky would go to her. Lucky went to see if Carol had crackers. It was good footage. Carol petted and sketched the yearling while the film was rolling. A young buck I had raised two years before came over to see what was going on. His small nubs of growing antlers were still in velvet. He wanted the crackers also. The videographer took more footage. Then we ran out of goodies for the deer.

THE ZOOKEEPER'S DAUGHTER

The two young deer stayed right by Carol expecting another hand out. When none was forth coming, the buck decided to look for himself. He started to nibble on the strap of Carol's art/camera bag. Carol gently pulled the strap out of his mouth, knowing the film inside the bag was poisonous. The young buck thought this was the beginning of great fun. Young bucks love to play push me pull me, an adolescent deer equivalent of slap and tickle. Immediately the young buck pushed Carol's bag with his knobby head at the same time pawing her arm with his sharp black hoof. I realized this could be the beginning of trouble, the same time Carol did. Carol immediately stood up; the videotographer dropped his camera and ran to Carol's aid. I pushed the buck out of the way, clapping my hands loudly, saying with my mean mommy voice, "Go on, git!" The young buck backed off. We decided that discretion was the better point of valor, we should not play a push me-pull me tug of war with a buck. The videotographer picked up his camera, we dusted ourselves off and slowly (so as not to encourage a chase game) walked out of the deer run. *Just my luck to take a great artist in the deer run and have her arm whacked by a boyishly playful buck.*

Years of experience with animals have taught both Carol and I to avoid potential disasters. I breathed a sigh of relief, as no one was hurt. We chuckled afterward when we discussed what had happened. By pulling the camera strap out of the mouth of the buck, Carol had inadvertently triggered the beginning hormonal instincts of the young buck, the push me-pull me trigger for a fight. A caring human response had started a whole 'nother response in the buck. Live and learn.

Chapter 79
Hope and Hannah, African Atlas Lions

Hope and Hannah were born on July 27th. Another two little female lions that the mama did not want, *Come on now*, again? I was home in the morning getting ready to go to work when Doug called and told me the cubs were on my desk. Only two in the litter, I expected them to be larger than they were. Hope was barely a pound, hence the name Hope, as in I Hope she makes it. I had lost the only other lion cub that was born under a pound. Hannah was larger, almost two pounds. After two days Hope caught up to Hannah in weight. She was drinking as much of the body building formula as her sister was, two ounces at a time, every two hours around the clock. When the cubs were the same size it was hard to tell them apart. Their soft tan and dark spotted coloration was very similar. One day in desperation I grabbed a blue permanent marker and dabbed a spot on the slightly larger cubs paw or hand. From that mark on the "hand," came the name Hannah.

The tiny cubs were constantly with me. They went to the zoo in the daytime, home at night, and sometimes sat in at meetings I could not avoid. I took them to Dr. Brian Voynick's TV show *The Pet Stop* on News 12 New Jersey. Again my germaphobia kicked in and I would not let anyone touch them, except the vets and myself. Hope and Hannah won a New Jersey's Best Ribbon from The New Jersey State Fair®/Sussex County Farm and Horse Show. They were the *only* lion cubs at the fair but, still, the ribbon was nice.

I loved the little girls; they were all that lioness cubs should be. My girls were more affectionate and social than male cubs. They would look at me while I was feeding them with love in their eyes. This is a look that is hard to resist. These two girls were born late in the zoo season. In the spring I am loaded down with babies in need of human intervention. I really enjoy the cubs that I have more time to play with. I started Hope and Hannah on ground meat at four weeks. Right on schedule at seven weeks old, they started to attack the bottle. Every cub I've raised hits the seven-week mark and starts to get aggressively possessive of their bottle. Perhaps this is practice for the future kills.

Tuesdays are the Day family's day off. First thing in the morning I go to the zoo and feed and clean the nursery. Then I come home to do the normal housework that every working woman does. The zoo's kitchen staff or my dad covers for one or two feedings during the day once all the species in the nursery are drinking well on a bottle. In the Springtime I do all the feedings myself. I'm usually back to the zoo by closing for the night feedings.

The tragedy of 9/11 effected us all. It will be a day that every American remembers what we were doing. At 8:30 a.m. I had just gotten home from my morning chores. Relaxing for a moment with my coffee, I watched the terror begin on TV. I, like all Americans, was appalled. Doug and I canceled our day's activities and stayed by the TV. I went back to the zoo and fed the nursery at 11 a.m. and 2 p.m. The animals had no idea what had happened, they all still needed to eat. I spent a lot more time playing and loving Hope and Hannah that day, and waiting for Jackie to come home from school.

The word went out to the other zoos, sanctuaries and trainers that we had two little female cubs available. I waited for the inevitable phone call. I picked a private trainer from Wisconsin. She had all her paperwork and asked all the right questions to pass my test. I would drive the girls to the airport on October 27th.

Airport security had been tightened. I needed to take two crates, one with the lions in and one to be inspected for transport. I had all the paperwork in order to air ship Hope and Hannah. Dad made a transport box according to USDA specifications. The paperwork was taped to the top of the transport crate. I taped the nipples to the bottles in a Ziploc on top of the crate along with enough powdered formula for a couple of days. The new mom/trainer on the other end had ordered the formula but it wouldn't be in for a couple of days. Time came and I steeled myself for the drive to the airport. My mom rode with me for emotional support. *47 years old and I still want my mother to come with me. Mature, Lori, very mature.*

Hope and Hannah slept all the way. When we arrived at the airport I took the cubs in first. In order to show the airport personnel how great the cubs were, I fed them on the floor of the cargo office. The airport personnel fell in love with my babies according to my plan. I figure that if the airport people can interact with my babies they will keep a special eye on them until they go. I'm sure they treat all the animals well, but I feel better if I know someone will keep a special look out for their well being. Those cargo hangers are big, noisy and scary to me and I am sure they are to the cubs, too. Feeding the cubs insures that they will sleep most of the way. The girls had a 5-hour flight. That would not be a problem.

I left Hope and Hannah in the care of the security lady, while I walked out to the car to get the transport crate. Security needed to check it for bombs. Ok, no problem with me, I wanted my babies to arrive safely along with everyone else on the plane. I was wheeling the transport crate on a cart up the ramp when I heard a sharp voice.

"STOP!" I looked at the security lady who two minutes before helped me feed the cubs. "What is that?" she questioned, pointing to the Ziploc full of white powder.

"It's lion formula." I shrugged, not quite catching on to the problem.

"You cannot ship that," she stated authoritatively.

" The crate? It's standard USDA approved." I still had no clue what the problem was.

"No, the powder," the suddenly stern security lady answered.

Then I caught on. *God are you dumb, Lori, with anthrax terrorism still happening you try to ship a Ziploc full of white powder on an airplane!* I looked around to see if the whistles and sirens were for me. *Whew, only my imagination.* "It is lion milk formula. I can eat it if you want?" I queried. I opened the bag and stuck my finger in.

"No, you just can't ship it," the security lady replied.

"OK." I shrugged. What could I do? I ripped the duct tape off the Ziploc and gave my mom the powdered milk to take back to the car. Security checked the crate thoroughly and it passed. I said goodbye to the girls. *At least they have each other for the trip.* I put the playful cubs into the crate. My girls were wheeled away.

"Don't look," Mom reminded me as we walked to the car. I couldn't help it, I looked. Both girls, two sets of golden baby eyes, looked at me. *At least they have each other for company for the trip.* I was shook up with all the extra stress. Mom and I headed home. The white-knuckle drive wasn't so bad this time. I didn't want to be at the airport. Saddened that my girls were leaving me, I was still relieved to be out of the airport.

Hope and Hannah arrived safely and were playing with the new mom/trainer when she called me. She loved the cubs. The cubs took an instant shine to her. The new mom sent me pictures of the cubs and herself the next week. She looks just like my daughter, Jackie. No wonder the cubs love her.

Chapter 80
Instinct

Webster's dictionary defines instinct as: The complex and normal tendency or response of a given species to act in ways essential to its existence, development or survival. Biologists define instinct as any form of innate (to be born with) behavior that is independent of the animal's environment, a non-learned behavior that follows an inherited response to certain stimuli. Simply put an instinct is an animal's performance of a characteristic pattern of activity that has not been taught.

There are three parts of an instinctual behavior: appetitive behavior, innate releasing mechanism and the consummatory act. Appetitive behavior is the build up or readiness for the instinctive act. This phase is goal directed, concerned only with the performance of the act. Innate releasing mechanism means that the animal has no inhibition (no reason not to) to perform the behavior. The final consummatory act is the relief of the animal now that the activity is complete.

Take, for example, a baby bird in a nest. The chick is hungry, food is waved in front of the chick, chick is in condition to respond because he is hungry (appetitive behavior). The chick opens its mouth wide to receive the moving food, eagerly, without hesitation or reservation (innate releasing mechanism). The chick lunges, takes the food in its mouth, swallows and is satisfied (consummatory act).

Another example is in the running of a herd. The running of a herd of animals is instinctual. The herd is in condition to respond if running due to fear (appetitive behavior). One member of the herd gets scared and starts to run, causing the rest of the herd to run also—they don't know why they are running, I'm running because you are (innate releasing mechanism). The consummatory act is the running of the herd. It makes them feel good (safe) to be running. Then the herd stops.

All instinctive behaviors are started by a sign stimuli. A sign stimuli is the environmental or internal reason for the beginning of an instinctual behavior.

The sign stimulus in the chick is hunger. The sign stimulus in the running herd is fear (or sometimes pleasure) of an individual. Other instinctual behaviors may be stimulated by hormones (mating, nesting, protection of young, feeding of young, fighting for supremacy of the herd), fear (an animal will fight or flight), protection from weather, or lack of food. Hormones, hunger and fear are examples of internal stimuli. Protection from weather, fear or lack of food sources are environmental stimuli. The pulling of the camera strap out of the buck's mouth was an environmental stimulus of a hormonal instinct.

The fight or flight instinct causes an animal to decide instantly if it can out-fight an adversary or if it should run away. For example, a rooster will fight anything its own size or smaller—another rooster, small children, even raccoons. Faced with a large adversary, the roosters immediately try to flee. Any animal will run if chased by something scary (bigger, meaner, or nastier than itself). Some animals will chase if they observe running. Our native black bear, your pet dog, and the tigers (in the enclosure of course) at the zoo are good examples. They run after you if you run. If you stand your ground they stand also. This is also part of the flight or fight instinct. If you are confronted by a native black bear, do not run, but walk to safety. Protection from the weather stimuli causes certain animals to hibernate, like woodchucks, others to migrate, for example, the migratory birds. Availability of food stimuli causes the animal to move where the food is.

There is a difference between instinct and learned behavior. Learned behavior is just that, behavior that is learned from experience or taught by a superior. You are hungry (an internal stimulus, appetitive behavior), you want to eat (innate releasing mechanism), you eat (consummatory act). When you go to the refrigerator to find the food, that is a learned behavior—someone taught you. You know the food is in the fridge because you have experienced that before. If a predator is hungry it seeks out food (appetitive behavior), it wants to eat (innate releasing mechanism) has to find and kill its own food and does so without hesitation, and eats (consummatory act). The difference between the two is that we humans have someone else find and kill our food (the butcher) and put it in the refrigerator. We "civilized" humans are only one step away (learned behavior) from finding and killing our own food by instinct. *And you thought that steak grew on the Styrofoam plate in the supermarket?*

A spider weaves its web by instinct (internal stimuli). If the web is broken by the environment (external stimuli) it will be repaired or strengthened by learned behavior. A young robin can build a nest, but an older robin learns to build a better nest. The nest building is instinctual, the improvements are learned.

A mouse is not afraid of a snake by instinct. The mouse does not know the snake is bigger than it is. It sees something the same size (the snake's head) the snake seeks out the mouse by instinct due to an internal stimulus (hunger). If a snake has chased a mouse, it is instinctual for the mouse to run (flight or fight). If the mouse escapes, it has a learned experience behavior of fear and will avoid the snake. It could theoretically pass on this learned behavior to its young if the mouse was smart enough. However few mice have the experience of escaping a snake, therefore do not have the learned experience to teach its young (learned behavior). The mouse is ignorant of the danger, therefore does not fear until it is chased stimulating the fight or flight instinct.

A shadow passing overhead will cause an instinctual reaction to animals including humans. This external stimuli causes fear. It is dangerous to pet a strange dog on the head. The extension of the hand above the dog's head, or a human head for that matter, causes fear of a strike. Always ask the owner first if you can pet the dog. If you let the dog smell your hand first, he will know you mean no harm. I often caution kids at the zoo not to pet an animal on the head first. Let the sheep, fawn, calf, or whatever smell your hand first, then pet under the chin or on the side of the animal. This is animal "Howdy do" common courtesy. Humans prefer to shake hands in the United States, other countries have other customs. Try waving over somebody's head and see what reaction you get.

Animal instinct and human instincts are very similar. We all have the same goal: to survive and reproduce. Once you understand that very elemental point, both animal and human normal behaviors are predictable. The animals I never worry about. They will do what comes naturally as long as they are not mistreated or sick. Knowing an animal's history and instinctual behaviors helps to deal with that animal on a daily basis. The same would apply to humans if you think about it.

Chapter 81
Swimming with the Dolphins

My daughter, Jackie, and I had always shared one dream: to swim with dolphins. We had watched commercials on TV for different possibilities and heard the reports of problems that may arise. I knew that swimming with wild dolphins is illegal. Doug and I made reservations for us to swim with the trained dolphins at a well-known resort in Florida. We had to make the reservations a year in advance. For that year Jackie and I were waiting in anticipation to put swimming with the dolphins in our life book of experiences. Thanksgiving week of 2001 was our vacation time. The three of us packed up the car and headed for Florida. I was psyched. I was ready for the experience. I had seen a wild moose in Canada, watched Loggerhead turtles lay eggs at midnight on at Juneo Beach Research Center, raised and interacted with many species of animals at home. But I have little experience with marine animals.

When the big day came we were the first people in the door. Our swim time with the dolphins was 10 a.m. Doug was nonchalant about the swim. Jackie, the all too hip teenager, was way too cool to show excitement. I, on the other hand, was unencumbered by the restrictions of "coolness." I was giddy with the excitement. It should be understood that at my age, I am no longer the bathing beauty of my youth and would not be caught dead in my bathing suit in public. Somehow my inhibitions evaporated with the breeze. I was a kid in a candy store.

The park itself was breathtaking. The sculptured lagoons and landscaping were so natural looking I could have sworn I was in the Keys, not inland Florida. *Oh to have that kind of money to fix up our zoo at home! We can all dream.* The trainers gave us a talk on Dolphins and their behavior. The trainer in charge put on a wonderful informative program about the dolphins in the lagoons. I responded to the trainer's enthusiasm with my own. Jackie kept poking me in the side, "Mom, shush." The large group of us "civilians" was divided up into groups of ten. We headed for the sandy lagoon.

THE ZOOKEEPER'S DAUGHTER

The weather was chilly for my perception of Florida and I would not be tempted to swim at that temperature normally. However I was undaunted and the first one in the water when the trainer said we could wade in. The sun was shining across the surface of the blue-green tinged water. *The filter system here must be extensive,* I thought.

In smaller family groups we were introduced to our trained dolphin. Jackie and I were the second group to go out to the dolphin. We followed the trainer's instructions to the letter. He told us which hand signals to use to communicate with the dolphin. The dolphin understood the first signals and came to us. We petted his wet, leathery gray skin. It had an unusual feel. The skin was warm to the touch, as dolphins are warm-blooded. Wet, but not slimy, we stroked the muscular side of the dolphin as he swam by us. The trainer signaled and the dolphin glided to the first couple treading water out farther in the deep lagoon. The dolphin splashed by them and took off to go see one of the other dolphins working with one of the other groups. Our trainer told us all to get out of the water. I was disappointed momentarily until the trainer explained.

Since the dolphin did not want to participate, the dolphin would not be stressed out and forced to play with us. And we would not play with him. I could understand that, only positive rewards work with larger animals at home. The group of ten of us stayed on shore discussing dolphins in general, their behaviors, body structures, breeding habits, birthing procedures and the mammary flap of the mothers.

The trainer asked for questions. The trainer fielded a few questions. Then I asked my question: "What is inside the mammary flap? Is it a flat mammary wall similar to the marsupial or are there one or two nipples inside the body?"

"MOOOMMM!" Jackie rolled her teenage eyes. I ignored her. The trainer answered that there is a flow of milk from the mammary walls inside the flap and the tongue of the baby dolphin actually rolls into a straw type shape to insert into the mammary flap as the milk flows out. *Neat-o*.

"You work with animals?" the trainer asked.

"Yes, I'm a zookeeper up north." I gave a simple reply as Jackie was giving me another one of her teenage looks.

The other three groups of people with dolphins in the lagoon were coming out of the water. *Darn, I wanted more time with the dolphin.* Then to my ultimate joy and surprise, the trainer announced we were going back in the water. *Yahoo! All right! Far Out!* In deference to my daughter and not to scare the dolphins, I walked as calmly as Jackie did to the edge of the lagoon.

It was Jackie's and my turn to swim out to the deep area. The trainer signaled to the other trainers in the lagoon and they all came over to us with the three other dolphins. We got to meet them all. Jackie and I gave the signal for dolphin kisses and had our pictures taken. We shook hands with the dolphins, and had them "swim" us back to the shallow by holding the dorsal fin as instructed by the trainer. I was euphoric with the heady experience of a lifetime wish come true. We met Doug on shore where he had been waiting for us. He could tell by the light in my eyes that I was thrilled with the experience. Doug asked Jackie what she thought. "It was all right, I guess," she replied. *Teenagers, ugh!*

The rest of the day was spent swimming with the rays and amongst the tropical fish in the lagoon. Doug enjoyed the aviary. The colorful tropical birds flocked to him and Jackie. I took the pictures. I also bought the pictures the park photographers took of us with the dolphins. The experience was worth every penny. I do not pretend to know a lot about marine animals, and I learned a lot that day. I would recommend it to anyone who loves marine animals.

Time was getting late and we needed to head back toward New Jersey and the animals we were more familiar with. We changed our clothes and climbed, exhausted, into our car. On the way out of the park Jackie turned to me grinning and said, "That was really cool, huh, Mom?" With just the three of us, she finally expressed her excitement.

Chapter 82
Holiday Fun in 2001

Hi!
First pony express, letters, then phone and fax,
portables, then cell phones, emails to the max.
So here's an old fashioned letter, sort of like me,
to hold in your hand, with coffee or tea.
Time flies fast, can it be a whole year
since my last letter of holiday cheer?
The year 2001, my life in review,
sending friendship and love directly to you.
Let's see, what's happened in my year..
now that I have your eye and your ear:
Got an article published in a magazine of the zoo trade,
"Bottle Breaking Multiple Species,"
Another banner in my parade.
It was pretty long, but hey what the #**#
It made me feel unbelievably swell.
Winter doldrums ended early, the first of March,
We cleaned all the dolls with lye, bleach and starch.
From the county paper an honor was given,
Named me one of the outstanding county women! (*Who Me*?)
To mention it seems quite vain, even so
I was extremely flattered, ya know?
By April three orphaned raccoons appeared,
with beady masked eyes, striped tails, and cute ears.
Two gray squirrels, a llama, and chickies galore,
duckies and Guinea fowl, peacocks and more.
Five fawns, one named Chester, I think he's a dwarf,
who may have trouble with the snows here up north.
Took Chester, the llama and coonies down to News12 TV,

they were a big hit with Dr. Brian, Dr. Ted and me.
A dear friend named Al passed away,
but gave me sweet "Fritzy" before his last day.
Fritzy, a girl black Australian Swan,
and Al's other two now grace our lawn.
Then in July, what I'd been waiting for,
two little girl lions, I'd hoped there'd be more.
Took the girls to the Fair, speeches and TV,
they grew fast and happily.
Did PR talks and llama walks.
Painted the facade of the building, way up high,
on a fork lift in a basket, thought I would die.
Got sea sick but hey, after all
no one can say I ain't got no #**#s!
(Sorry to those whom language can upset,
there's still some redneck in me yet.)
The zoo was busy, lines out the door
Kept me hopping, never a bore.
I'll take credit, where credit is due,
but some days are brutal here at the zoo.
Tara the tiger miscarried this fall,
I was saddened, but you can't save 'em all.
Took the lion girls to the airport in October
after 9-11 the mood was sober.
White powdered lion milk, sent in a baggie,
brought security quick - all looking at me.
Then I explained that it was milk for the cats
They would still not let me ship it, imagine that!
Doug's a great guy, he puts up with me,
and that's an amazing ability.
He designed a new winter room built in a barn,
to keep all our southern critters cozy and warm.
Jackie's a trip, changing every day,
as tall as me, but like Doug in some ways.
Cute as a button, smart as a whip,
some freshman classes she was able to skip.
She's always on the net, in case you call,
or trying to get me to drive to the mall.

THE ZOOKEEPER'S DAUGHTER

We went fishing this summer, Canada again.
This fall it was Florida, it was our yen.
One for our life book, we've wanted to do:
to swim with the dolphins - I've got pictures, too!
We played with the dolphins and got dolphin smooches,
they were trained very well, like water logged pooches.
Two more little llamas, this job never ends
they'll be Christmas presents for two happy friends.
So to all my friends both short and tall
Holiday Cheer to one and all.
Pony express, letter, email or fax,
I'd love to hear from you back.
Put a smile on your heart as you think of me,
while holding my letter with coffee or tea.
Love Doug, Jackie and Lori

Chapter 83
The Lions Finally Catch On

January 1st may be a holiday to the rest of the world, but here at the farm chores still need to be done. New Year's Day dawned late and unusual, a winter without snow. The brisk air whipped through the trees in the zoo while the animals all snuggled in their warm hay-stuffed dens. Doug and Parker were out early to feed and water all the animals. That was all that would get done on a winter holiday. Except, of course, Dad would make his rounds on the country roads picking up the dead deer that were killed by people out partying late the night before. I didn't go into work. The lion cubs had been shipped out months before and I was baby-less. I go through empty nest syndrome every now and then, too.

Doug came home from his feeding chores with music to my ears. The lioness had cubs, and was in the den with them. Parker had seen two, both were nursing. Solomon and the other lioness were also in the den and being protective. *Finally!* I knew the cubs would be kept warm by the mother's side, we were having a mild winter. As is our procedure, we kept an eye on the mom and cubs. She was taking good care of them, protective and snarling when the men came near. We decided to leave the cubs with the mom. Our old male, Solomon, is becoming ancient, gray in the muzzle, skin wrinkling like the old man he is. We hoped one of the cubs would be male. The magnificent black mane of the Atlas African lion is carried genetically by the females but only expressed in the huge masculine lions. Between Solomon's heavy black mane and the cubs' maternal grandfather's extensive mane, from head to loins, a male cub would grow into a wonderful specimen. We could only watch and wait.

Doug and Parker change the hay in the big cats' dens every Friday. The momma lion would not leave the cubs. That was good maternal behavior. The momma and her sister and Solomon were all protective. All good signs. The bad news was the hay would not get changed. Luckily there had been no snow for the cats to track in and the hay was still dry. Not wanting to force the mother out, the guys skipped haying the lions' den that week. The following week the momma lion was still taking great care of the cubs. Our hopes were high that this time she would raise the cubs and the line of Space Farms magnificent huge Atlas lions would continue. On Friday the guys once again went in to see if they could change the hay. Both mommas said no, snarling and coming at the men. When the mommas charged the gate, Parker saw three more newborn cubs. This, as you recall, has happened with our females before, having litters within a week or so of each other. The men left the den undisturbed.

Dad, Parker, Doug and I discussed the situation. We did not need eight lions. I had other zoos, trainers and sanctuaries on my waiting list. It was decided to leave the cubs with the mommas till Monday when we would try to remove the smaller three. Over the weekend I contacted Dr. Spinks to get the vaccination shots for the larger two. If we could get the mothers out, this would be a good chance to give the larger cubs their first shots. The next inoculation shots would not be so easy as the cubs raised by the mother would be "wild." That means just about un-handleable by eight weeks old.

Monday came and we gathered our tools. Dad had plan A for getting the cubs out and Parker had plan B. We would try Dad's plan first. Crawling on all fours against the wall of the three-foot walkway in the den, 73-year-old Dad made his way between the internal bars and the outside wall of the den. The lower section of the bars were closer together so the lions could not reach out and grab a person. The upper bars were farther apart and a paw could come out. My 73-year-old dad crawled down the walkway with a set of bolt cutters to cut a section of the lower bars. That was the corner the cubs were cuddled in. The theory was the cubs would crawl out. Plan A did not work. The females were so upset. It was decided to let them settle down for a while.

After lunch, we tried plan B. Doug locked the tigers (Khyber and Tara are very cooperative) out on the grassy section of their compound. Parker unlocked the tiger den gate and walked out of the tiger den door to the concrete section. He calmly took off his trademark baseball style cap and flashed it at the momma lions inside the lions' den next door. The lionesses charged out the den door toward him. Parker has nerves of steel. The carnassial teeth of a lion are six inches long out of the gums, and grow in the skull from a point under the eye socket. These females were roaring and charging the partition between the tigers and lions' compound. Doug's job was to drop the heavy steel slide on the lions' den when the lions were outside. Doug grabbed the thick cable and the steel slide slipped into place—on the tiger's den. OOPS... Wrong cable.

Doug got the cables untangled and Parker tried again to entice the lionesses outside. The snarling 400 lb. females came out to challenge Parker once again. One of the larger cubs followed the mommas out and was directly under the slide. I hollered to Doug to hold up the slide. I had the sight vantage point and could see the cub. Parker swung his hat against the dividing bars again. The lionesses stayed to challenge Parker-the-intruder and the little scared cub ran back inside. Doug successfully dropped the slide. The cubs were inside and the lionesses were out. And boy were those ladies angry!

I was prepared with a box and toweling for the cubs. The guys handed me out four cubs, two large, healthy butterballs and two smaller cubs. "The third cub must be gone," Parker said. Often in the wild the small cubs will be pushed away from the nipple by the stronger cubs. Dead cubs are eaten by the mother. We knew the second litter would not make it if we left them in the den. While I was giving the two older, chunky cubs their shots the guys pitch forked out the old hay. I checked the cubs' sex, two males and two females. One set in each litter. When I was done, I helped with the hay removal. New fresh fluffy hay was placed in the den and the older, fat-on-mother's-milk cubs were placed back on the dry hay bedding. We expected the cubs to be wilder, but at three weeks old they were still very passive. Gates and doors were double-checked for chains and padlocks. Doug hoisted the cable for the steel slide door. Both mommas rushed back in to the cubs.

Dad came back after his dead deer pick up and congratulated us on a job well done. I'll tell you, folks, I never want to stare down a charging angry lion. My brother has intestinal fortitude. I took the two cubs home to set them up with bed and bottles. I knew that both the mommas would nurse the two remaining babies, that behavior is common in the wild. Both mommas were attentive to the remaining cubs. Since the cubs were left with the mommas, the mommas will not breed again for at least a year, in theory. That made the babies in my house more special, these would be my last lion cubs for a while. I thoroughly enjoyed them. I found good homes for both, the new parents asked all the right questions, had all the legally correct paperwork, and said that the cubs would work with celebrities and on TV.

By the end of February, Mom and I were once again making the dreaded trip to the airport. I've made the trip many times, but this time I took the wrong exit and ended up in downtown Newark. Thank God Mom was with me and had grown up in that area. She knew which streets to take. We arrived at the airport and set the male up to go to Texas, the female to Florida in the wooden transport boxes Dad had made. It's always hard to say goodbye and not look back. I knew these would be the last little lions for a while. I had to say goodbye twice as the cubs were flying out on different airlines. I still looked back, but this time, each cub was sleeping with a full belly ready to fly away. An omen they would be fine. Mom and I drove home.

    I received phone calls later that night. Two new adopted parents in different states were thrilled with the new babies. I slept well. I saw little Biscuit, the female, on TV in early March with a well-known animal handler. Her new mom was taking great care of her, I could tell. Biscuit looked happy and had that loving glow in her eyes. Her life will be good.

Chapter 84
Little Boy Blue in the Nursery

Every spring thousands of Americans buy duckies or chickies for their kids at Easter. In New Jersey you must buy six or more at a time. This helps eliminate the average Joe in an apartment from buying tiny cute fowl, only to have them grow up to be large, messy birds that he inevitably wants to get rid of. It is a great law, but I think it should be expanded to twelve chickies or duckies.

Every year without fail, someone calls the zoo wanting to donate his or her now full-grown ducks or chickens. We usually turn them down, as there is a limit to how many fowl our pond can carry. Our pond is also visited by native fowl that fly in and eat the homegrown corn we put out for the farm fowl or the fish that naturally inhabit the pond. Great Blue Herons, Mallard ducks, wood ducks, coots, mute swans, osprey, eagles and many other birds frequent the pond at the zoo. A bird's paradise to be sure, but we have to limit the number of farm fowl to encourage the wild birds to come in.

I did, to my chagrin, take in six puddle ducklings in the spring of 2002. The tiny, fuzzy, sunny yellow balls of fluff looked so cute in my nursery. I set them up with a kiddy pool inside the nursery fence. The sunny six would waddle around on the grass quacking to the visitors, seeking the food the visitors would throw. After their tummies were full, the sunny six would tumble into the kiddy pool to paddle around until their next parade into the grass. They stayed safely within the nursery fencing.

Also in the nursery that year were five little Mouflon sheep. The Mouflon sheep had come from our herd and were being raised on a bottle to go to other zoos as part of a genetic diversity program. This also helped thin our herd of Mouflons as it was getting quite large. The lambs' tan bodies with black stockings made them interesting creatures. Keeping the lambs company were two Spanish La Mancha goat kids. This type of Spanish goat is specially

THE ZOOKEEPER'S DAUGHTER

bred to have no ears. In Spain the goatherds had a problem with ear parasites and bred a genetically recessive breed of goat whose ears were little more than belly buttons on the sides of their heads. These kids certainly looked strange and unusual. They could hear perfectly well, the only drawback to the special breeding is that the goats could never wear sunglasses! The duckies, lambs and kids had free roam of the inside of the nursery. Two coyote pups were inside an enclosure and a dozen gray, white polka dotted Guinea fowl inhabited the other cage inside the nursery. My nursery was chock full and very active with the duckies parading amongst the prancing kid goats and lambs. And very noisy with all their talking.

Every morning the nursery was fed and cleaned. The lambs and kid goats received bottles four times a day. The sunny six duckies would get their turquoise kiddy pool, with blue and white duckies printed on it, cleaned and filled with fresh water for the day. The Guinea fowl would be fed and watered. Often the coyote pups would squeeze out the enclosure door and chase the kid goats and lambs around in play, practicing their hunting techniques. Somehow I always managed to catch them before any harm was done.

My biggest problem in the mornings was feeding the seven hungry mouths of the kid goats and the lambs. They all wanted fed at the same time. Over the years I have developed the talent of feeding five hoof stock at a time, holding two bottles in each hand and one bottle squeezed tightly between my legs. I know it looks strange, but it gets the job done. Seven hungry mouths I simply cannot feed by myself. Often I would let two out into the zoo proper and feed five, then call the other two back into the nursery and feed them separately. Or I would enlist any other person I could find to help. Once the electric company meter reader was there and she helped. The morning feedings were difficult, to say the least. The mid-morning, mid-afternoon and six p.m. feedings were easier, there were always visitors willing to help.

In the afternoons I would announce the nursery feedings over the public address system and have all the help I could use, usually more than I could use. I like to give children the chance to feed the bottles. It was fun for the children and a good photo opportunity for the parents, not to mention good public relations. I would give a short speech on the animals in the nursery as the human children fed the kids and lambs.

One Sunday afternoon I had about a hundred people and children waiting to help me feed bottles. I chose the human children by birthdays (the closest birthdays to the day) and sorted the seven children out by size. The smaller children were to feed the smallest lambs, the bigger children were to feed the kid goats. I explained to the children how to hold the bottles above their heads so the babies could not get the bottles until we were ready. I would give the count of three and the children were to put the bottles next to their own belly buttons pointing to the mouths of the babies. I knew from experience that if one human child put his bottle down first all seven hungry mouths would converge on that one bottle inevitability scaring that child. I also told them to hold on tight as I knew the lambs and kids would come running fast and "buck" for the bottles. The children were excited and the parents ready with their cameras. I opened the gate. The human children filed in quickly, as I had instructed. The lambs and kids charged the human child line looking for their afternoon bottles. I counted to three after the human children were lined up facing their parents for the great pictures to come. During the count of three one of the lambs had jumped its front feet up on a smaller child and knocked her down. I hurried to her aid, setting the child on her feet, helping with the bottle. I checked the rest of the line. The other children were holding on tight to their bottles, lambs and kids nursing away with their tails wagging as they pushed harder on the bottles. I started my speech on the animals in the nursery. The parents clicked their cameras. The ducklings were waddling amongst the feet of the children, lambs and goats, looking for food. Parents, grandparents and siblings outside the nursery fence were shouting instructions to the children feeding bottles for better pictures. The noise level was hard to speak over.

Suddenly, out of the corner of my eye, I saw movement. The black and white kid goat was pushing and bucking on the bottle held by one of the larger children on the opposite side of the line. The little boy was about eight years old, wearing sweat pants and a blue jacket. His big brown eyes were growing larger as the goat nursed more heartily. The little Blue Boy was slowly backing up, backing up toward the turquoise kiddy pool filled with water for the sunny six ducklings. I could see what was going to happen, and so could every parent outside the nursery fence. Everyone started shouting at once, making any one voice impossible to hear. I quickly stepped toward the moving child, my hand extended to catch his coat. I felt the wisp of his jacket as gravity took over. In the blink of an eye, the little boy fell butt first into the kiddy pool filled with (slightly) used duckie water. The sunny six heard the inviting splash of water and ran back to join the little boy sitting in the cool water. The kid goat was still nosing around for his bottle. The surrounding crowd burst into laughter and I must admit, so did I. I saw the wet child tear up and turned away so he could not see me laugh. But I am sure he heard the crowd. Once I had myself under control I helped the cold Blue Boy out of the duckie pool. My heart went out to the boy with the big, brown, teary eyes. I asked the crowd for a round of applause for my assistants, pointing to the children still holding bottles in the line. Then I asked for a special round of applause for the wet boy bedside me and asked the crowd, "Don't you think he deserves a free Space Farms t-shirt?". The crowd cheered. I put my arm around the boy and told him and his mother to meet me in the gift shop for a t-shirt and that I would get him a jacket out of lost and found. He looked to his mother and spoke a few words of Spanish, and she replied in a soothing voice, but what she said, I did not understand. Mom ushered the Blue Boy toward the building. I figured I would meet them inside. I escorted the other children out of the nursery and collected my bottles. I went inside to find the little Blue Boy and his mom. I could not find them. They had headed straight out the front door. The boy never understood my English instructions. I owe that little boy a t-shirt for being brave.

Chapter 85
Angie, the Mollucan Cockatoo

Our Zoo has a long history of surprise donations waiting on the doorstep in the mornings. I'm not sure if this happens at other zoos, maybe our visitors know what soft touches Dad and I are for animals that need help. We do not like to receive animals this way, we need special information on the history and medical conditions of animals. I knew something was up when Dad called me at home at seven in the morning.

"Are you expecting anything?" were the first words out of his mouth. Not, "Good Morning," not, "Guess what we got?" just those four words.

"Not that I know of," I replied, placing my coffee on the counter, as I finished lacing up my sneakers. I was out late the night before and I was only a few minutes late for work.

"You better get down here, you have a surprise." Dad laughed.

"Lion cubs?" was my natural response, although we were not expecting any. "Tiger cubs?" was my next guess. I was hoping for tiger cubs.

"Nope, just get down here and see."

It was early August; the weather was warm and balmy that time of the morning. I immediately left for the short drive to the zoo's main building.

As I pulled into the parking lot I saw Dad standing in front of a three-feet square black cage sitting on the sidewalk by the front door. Dad's round body blocked my view of the cage's interior. He was doing that on purpose, I could tell by the ****-eating grin on his face. Dad was holding a red file folder in his hand. I quickly walked toward the surprise, which was at this time making a whistling noise. I knew at that point it was a bird of some sort. Imagine my astonishment when Dad finally stepped aside and I saw what was inside the mysterious cage. A full-grown Mollucan Cockatoo peered up at me with soulful black eyes.

"Where did he come from?" I asked Dad.

"Don't know. Here's the note that was on top of the cage when I came in this morning," was his simple reply.

I scanned the note. "Angie" kept his eyes on me all the while, whistling. The note was short, telling me the bird's name, and dietary preferences. The owner had gone into a nursing home, and the adult child lived in an apartment and could not have pets. These people had seen me on TV and hoped we would give Angie a good home. The note was simply signed, SD. A bag of food and a Tupperware of cooked noodles sat alongside the cage. This is not the way we like to receive animals. No medical history, age or description of disposition was included. We had a bird on our step that was capable of breaking a broomstick with its beak and no knowledge of how handleable it was. Or if it would get along with my other large parrot, Inca the Macaw.

I looked at Dad; he shrugged, and said, "Guess ya got another bird."

"Yep, looks like it," I replied. I took Angie into the office, changed the paper in his cage through the sliding tray on the bottom and gave him fresh food and water. I had no idea how friendly (or not) this large, salmon-colored bird was. He certainly had a large, black beak. I talked to Angie all through the procedure to introduce myself. Angie cocked his head to the side, observing me as he climbed the side of his too small cage (in my opinion). After making sure the new bird was settled in the office I went out to the nursery to complete my morning chores.

At the ten o'clock coffee break, Dad, Doug and I discussed what to do with the bird. Doug jumped on his new project that afternoon. Doug divided Inca's cage in half with a large piece of tin. We could not use wire, as we were not sure how friendly the two birds would be. There is nothing worse than a bird fight, if the two human-raised parrots did not like each other. The tin would prevent the birds from biting each other's toes off or fighting through wire. Inca's large flight enclosure, five feet by five feet by eight feet was divided in half with Dad helping Doug on the reconstruction.

I sought out two natural branches for perches. The captive beavers at the zoo do a wonderful job of de-barking the branches we feed them. The de-barked branches are taken out of the beavers' enclosure to prevent them from making a lodge and hiding all day inside the lodge. I picked out two long de-barked branches of appropriate diameter and scrubbed them with Clorox to disinfect any germs that might be on the wood. I then fastened them inside Angie's half of the flight cage. Water and food dishes were secured inside the enclosure so the bird could not push them out of my reach.

By the next morning the divided enclosure was ready. I carried Angie's small black cage to the door of the large flight cage and opened both doors. I remembered my pet shop days and made sure the outside door to the room where Inca and Angie would live was closed. I did not know if Angie could fly. I know Inca can. Angie hopped into the new enclosure quickly, appreciating the extra roominess. I observed the two birds side by side for a while. Inca knew something was up and kept calling, "Hello, Hello, Hey Ma, Hello, Hello." Angie whistled a tune and settled onto a perch surveying the front half of the zoo from his hilltop perspective. All seemed to go well, so I left the birds to get on with my zoo work.

Mollucan Cockatoos are large 20-inch tall birds with distinguished head crests. The Mollucan cockatoo is a pale salmon color with deeper coloration on the inside of the head crest. Males have a slightly darker coloration in the crest. Gray, scaly feet with black claws and large strong black beaks make the contrasting peachy color more vibrant. Mollucan cockatoos are also "chalky" meaning they give off a lot of natural dust and dander. The males have black irises and the females have deep reddish brown irises of the eye. This is the only external sexual identification mark. Wild cockatoos flock in large numbers and are very social and vocal. The Mollucan cockatoo is native to the South Mollucan Islands including Caram and Ambonia in Indonesia.

Angie did well all summer on the top of the hill with Inca as his condo neighbor. As fall approached, we put heat lamps over both birds to keep them in a tropical temperature. Macaws can take temperatures down to 40 degrees, but I was not sure about Mollucans, so better to err on the side of caution. The zoo closed for the season on schedule, October 31st. The first of my winterizing jobs is to bring in the two tropical birds. I had set up Inca's winter cage, which is large enough for me to walk into, about the size of two telephone booths stuck together. (You know, the kind Superman used to change his clothes?) I called Inca out and he came over to me. I grabbed him and wrapped him in my coat for the annual chilly trip to the restaurant where the birds live for the winter. We did not have another large cage for Angie. Doug would need to build him one as soon as the rest of the zoo was bedded down for winter. I figured Angie would be right at home in his old black cage and would just walk in. Well... I was wrong, again. Angie liked the bigger flight cage he shared with Inca and decided to stay inside. I put Angie's favorite food in the back of the smaller black cage and the black cage door to face the flight cage door. I stood back and waited for Angie to get his favorite snack. No go. I hesitated to grab Angie with heavy gloves on, as I knew he could bite through them if he wanted to. I still had no idea how friendly the giant pink parrot was, or was not. The day was waning on, the cold night approaching. I picked up a broom and shoved the handle in between the wires, hoping Angie would hop on and I could place him into the black cage. No go. The last option, before grabbing Angie with gloves and risking a finger, was to scare him into the black parrot cage. That worked like a charm. I swung the broom handle above his head and Angie trotted right into the black cage. I quickly closed the doors, wrapped the cage in a blanket and transported Angie down to the restaurant where Inca was waiting. I made a mental note that Angie was afraid of a broom handle swung over his head. Come to think of it, I would be, too.

Once in the restaurant, I set Angie up on a table next to Inca's large cage. The birds could keep each other company. I slowly reached into the back of Angie's black cage to pull out the food I had placed there as bait. I did not want to reach fast and scare an already scared bird (with that big beak). You can imaging my surprise when Angie climbed right up on my arm and headed up toward my face. I froze like a statue, knowing that any movement might frighten the strange bird and I may lose an ear, nose, eye, ah well, take your pick! I was still standing like the Statue of Liberty as Dad walked in the front door a few minutes later. Angie had crept up my shoulder and was cuddling the side of my head.

"Uhhhh, Dad?" I queried, looking for help. Dad saw the worried look on my face and immediately started talking to Angie.

"That's a nice bird, nice birdy," Dad crooned to Angie. *Great help he is, the bird is now rubbing my head with his big (and growing bigger) shiny black beak.* Dad extended his arm to the cockatoo and Angie climbed right on. This was a well-trained bird! We had no idea, as Angie had not shown any evidence of this behavior for the previous three months. Dad slowly moved Angie toward his cage, Angie hopped right on the top. Next I extended my arm and the marvelous bird came onto my arm and headed up my shoulder again to cuddle against my head. *Guess I'm all right.* I reached up my sacrificeable left hand and petted Angie on the back.

"Looks like you have another pet bird." Dad laughed.

Life is full of surprises. Dad took a moment and we had the affectionate cockatoo climb back and forth between us, just for practice. Then we let Angie stay on the top of his cage for the rest of the day. His cage was too small for full time internment. The end of that afternoon, I extended my arm to Angie, saying, "Come," and Angie ran over to climb up and give me kisses. He never put his beak on me, just made the sucking noise of a kiss. I smiled all the way home. I told Doug that night at supper that he would not have to make another big cage, Angie would be out every day and I would put him to bed inside his cage at dusk. Even though Angie has the run of the closed restaurant, he seldom leaves the top of his cage or the special perch Dad made for him. Angie has never flown for me; it could be his wing muscles have atrophied from many years inside a cage. Angie has not said any words yet, but has a whistle vocabulary of a number of songs.

    The next day as I cleaned the cage, Angie was on my arm again. In the winter the restaurant is very lonely with no visitors except for Dad, Parker, Doug and I. I often play the radio with my favorite oldie station (not that I am old). "This Magic Moment" by the Drifters came on the radio. It is a favorite song of mine, so, as there was no one else in the building, I started singing. Angie started to dance up and down and swing his head like Stevie Wonder. Wow, finally somebody liked my singing! I thought it might be just a fluke so when the next song came on, I sang that one, too. Angie started to dance again. The next time Dad came through I showed him my discovery. I was so thrilled, I thought Angie's dancing was really neat-o. Dad laughed and we sang a duet for the dancing pink parrot. I'm sure we looked quite funny. Dad was in his old, patched, farmer bibs and beat up, dead-deer-blood-stained, quilted flannel shirt. I wore jeans and a sweatshirt covered in paint. Angie was resplendent with his salmon coloring and his hot pink colored extended crest. The three of us were singing and dancing around an empty building to oldies on a staticy radio. I half expected the guys in the white coats to come and take us away! After that Dad would have me show all his friends how the bird danced. Everyone liked the dancing bird, but one day Dad joked with me that Angie was not dancing; he was convulsing due to my terrible singing. That was the last time I sang for Dad. But Angie still likes my singing. And I like his dancing, a regular mutual admiration society.

    Every morning Dad or myself would let Angie out of his cage. Hunting season (our High Holiday) came once again in December. As the men were having morning coffee one day Angie walked across the floor to where the men were standing and proceeded, according to Dad, to bite the boots of the hunters. I think Angie just wanted "UP" and was testing the boots for sturdiness before he attempted to climb up the legs to get the kisses he wanted. But that is my opinion. So whenever the men came in for coffee or lunch, Angie was placed temporarily back into his cage, to be released again after they left.

    One day as the men were eating lunch Angie was whistling up a storm, he wanted out to enjoy the party. My cousin Dutch runs a bus company and had been to the bank that morning to take out $450 in Christmas cash. He joined the hunters at lunchtime. Angie was really carrying on so Dutch swung his coat over the cage to quiet the noisy whistling bird. The peachy parrot quieted down, busying himself behind the jacket. The men gabbed about their hunting adventures and finished up lunch. When Dutch grabbed his coat to leave, Angie had a fifty dollar bill in his beak. In the bottom of Angie's cage, on top of his cage paper were two fists full of shredded fifty dollar bills! Needless to say, Dutch was angry. The entire hunting party roared with laughter. Dutch spent the rest of the afternoon scotch taping the torn bills together instead of hunting. It was quite fascinating; Angie had dexterously taken the crisp green bills out of the bank envelope in Dutch's inside jacket pocket. The bank envelope was still in the pocket, only the green bills had been removed. Every one of them. And the moral to this story? This bird needs more greens in its diet!

Chapter 86
The Boa of Stillwater

Early in September I was filling in at the admissions desk and took a phone call from the Stillwater police. Stillwater is not far from us and the police often call us on road killed deer so we can pick up the fresh meat for our carnivores. We also help all the local police with identification of the many wild snakes native to this section of New Jersey. This was a snake call, could we identify a large snake they had found crawling across a major highway? I chatted with the officer on the line and between the two of us we decided the 8-feet-long snake he was describing was not a native snake. It had to be an escaped or released pet. He asked me if we would take it in until the owner came forward. I hailed Dad from the restaurant, explained the situation, Dad said yes and I told the officer we would keep the snake for them. The officer was worried because the snake was snapping at the other officers. I instructed him on how to safely pick up the snake with a dog snare around its head and support the body as they lifted the large six-inch diameter snake into a garbage can with a lid on it for transport. This was a big snake to have wandering freely around town.

Two hours later the wayward snake was at the zoo with some peacock-proud strutting police officers. Dad, Parker and I went out to greet the officers and check out the snake. In the zoo business you accumulate all sorts of transport and temporary cages. We had an empty 70-gallon fish tank that the petulant pet snake was placed in. A large dish of water was provided for the snake. The new addition to the zoo had the nerve to snap at me as I placed the water dish in his tank. Dad and I examined the snake and told the officers it had not been out long, there were no tics or lice and the glistening snake had recently shed its skin. We temporarily identified it as a boa constrictor. The twinkle in Dad's eye as he examined the snake said it all. Dad was like a kid in the proverbial candy store looking at the wayward eight-feet-long pet snake.

Boa constrictors are non-poisonous snakes that squeeze their prey to kill it and then swallow it whole. Boas are ovoviviparous, meaning that the female retains the eggs within her body until they are hatched and born at the same time. There is, however, no umbilical cord supplying nutrients to the unborn. Baby boas are born at eight inches long and are often sold in pet shops as pets. They grow rapidly, given an adequate food supply. An adult boa constrictor can grow to be eight to thirteen feet long. In the wild boas live in Central and South America. There is one genera and many species of boa and many different subspecies, which are determined by color, size and geographic location. Your basic boa from a pet shop has beige, brown and black splotches of color. This was the coloration of the boa the officers had brought us. There are emerald green boas, red boas, boas that have a rainbow sheen to their scales and many other beautiful variations. Boas are bred in captivity specifically for the pet shop trade. Boas will climb trees in search of food, but are mainly terrestrial. Found in many types of habitat, the boas prefer dryer, warm environments, but will enjoy a good "soak" now and then, especially before shedding the skin.

My family has always enjoyed the diversity of nature and that includes snakes. Snakes do make good pets if they are handled and become tame, enjoying the warmth of the human skin. I would encourage anyone to have a pet snake if they take in all the information available on care, feeding and most importantly—life span and adult size. Always ask the person you are purchasing a snake from these important questions. And remember, the child you are purchasing this "gift" for will only stay at home for X amount of years and the snake may live way past that time. At the zoo we get at least one phone call a week on snakes that have outgrown the facilities or outlasted the interests of their human caregivers. We do not take in large, exotic snakes. Space Farms does not have enough winter quarters for the numbers and varieties of snakes the phone calls ask us to take in. It is a dream of Dad's to be able to do so, but that dream has not materialized yet.

Our wayward road boa was set up in the main building next to the restaurant. His tank and interesting story attracted a lot of attention. The local papers did a number of stories on the escaped pet. Dad and I took the eight-feet-long boa outside for pictures for one newspaper photographer. Dad estimated the weight of the snake to be a hefty thirty pounds. During the shoot, the boa constrictor snapped at me repeatedly, but seemed to love Dad, who has much more expertise handling large snakes than I. The patterned snake never snapped at him once. The photographer stayed way clear of the meandering snake's head.

No one came forward to claim her. There is a fine for releasing non indigenous snakes in the state of New Jersey, not to mention the endangering of children and pets that may come too close to a large snake. The bite of a large constrictor is not poisonous but can cause considerable bleeding and bruising, not to mention mental stress. One of my co-workers in the pet shop was nailed by a large snake. His hand was scratched by the teeth and remained bruised for weeks. Large, exotic snakes are very strong.

    I am glad to report the boa is doing well, spending its days coiled under a heat lamp this winter. She has eaten and shed her skin. Dad loves her, I think she has a permanent home.

Chapter 87
Kernel Corn
Does Not Grow Tomatoes

We sell shelled corn for our visitors to feed to the animals at the Space Farms Zoo. Feeding the animals is a big hit with the visitors. We supply the corn so we can control the quality of what the public gives the animals. Whole corn is a special sweet treat for our hoof stock and birds. But there are many kinds of corn that would give our animals major digestive and medical problems. Seed corn often has pesticides and fungicides painted on it. Popcorn is just about indigestible. If you are going to a zoo and want to feed the animals, call ahead and see what is or is not allowed. Each animal has special dietary requirements and or restrictions. If you are uninformed, you may inadvertently cause an animal harm. At our zoo we reserve the right to inspect food to be used for the animals and instruct visitors on the who, why and why not of who can get what.

America has a long history with the corn plant. Native Americans called it maize. It was cultivated by the eastern tribes and was a staple in their diets. Native Americans shared their corn and its growing techniques with the Pilgrims after the first Thanksgiving feast. Native Americans planted a few kernels with a scrap of fish head in the dirt mound for fertilizer. The new food was one of the first exports to Europe. "Indian" corn was multicolored and is used today for decorative uses. "Cow corn," is used for animal feed and has many different varieties. "Sweet corn," that we humans eat, also has many different varieties that you may be familiar with. Sweet corn has been developed for human food to be sweeter, i.e. higher sugar content, smaller, more tender kernels and also has been developed for color. Butter and sugar, white, and Silver Queen are a few common hybrids developed for human consumption. I can't tell you how many times the local farmers have chuckled at "city folk" who think they are getting away with something by cutting (stealing) cow corn grown next to the road, thinking it was sweet corn. It is not, which the city folk find out the hard way by the next day! Today, growing corn is a multi-billion-dollar industry. Amazing, it all starts with one tiny kernel.

Each kernel is a seed within itself. Planted in good earth, germination (starting to grow) will take place in 4 to 30 days dependant upon the warmth of the days and sufficient rain. Corn is planted in this section of the country in late April, early May after the fields have dried up from the winter snows and the weather warms up. Corn grows very fast in the right conditions. There is an old farmer's adage that "Corn should be knee high by the forth of July." Another adage says that you can stand in a cornfield and hear the corn stretching as it grows. There are many varieties of corn. Hybrids have been developed for larger, more kernelled ears, more ears per stalk, shorter stalks (so they do not blow over with a gust of wind) and disease resistance. Crop yield varies according to the weather and variety of corn. Some varieties yield 30 bushels per acre, others up to 100 bushels per acre.

As the corn grows the stalk elongates and the young ears develop. The silk (the strings coming out the top of the ear) is actually the conduit for the pollen from the tassel (the pine tree shaped spikey part at the extreme top of the stalk). When the wind blows the pollen is shaken off the tassel and falls gently onto the silk, which fertilizes the individual kernel inside the leafy wrapped cob. Only fertilized individual kernels grow plump and tasty with adequate water. Unfertilized kernels do not mature and show up as the dry nubbins we see on an ear of corn.

Farmers let the cow corn grow and wait until the first frost kills the plant, sending the sap back into the ground, in effect drying out the corn, leaving the nutrients in the kernel. The farmer then uses a corn picker, which takes the ears off the stalk and the kernels off of the ears. Often in the same process the stalks and the discarded cob are chopped into small pieces as fodder for bedding or to be plowed under for next year's fertilizer. If the farmer is making silage for his cows the corn stalk is cut and chopped green, ears and all. The chopped stalk is then placed in a silo and allowed to ferment. Corn squeezins' (the juice from this process) was often distilled by moonshiners for their bootleg trade in the old days.

In the old days before modern hybridization, a farmer could average 1,600 corn plants per acre. Nowadays with scientific hybridization, fertilizers, and large machinery, the professional growers can average 30,000 plants per acre. Weeds, insects and drought are the biggest problem for a corn crop.

Space Farms has 90 acres in corn crop, producing 300 ton of kernel corn for our zoo stock. We grind corn with the cob for our hoof stock and feed stalks also. Some of the corn we grow is stripped from the cob and sold as feed to our visitors, who enjoy tossing it to the animals and farm fowl.

One day a man was leaving the zoo with his very tired child. Dad and I were sitting by the zoo doors, conversing about the next day's activities.

"What can I do with this?" the departing visitor asked me, as he held up the plastic baggie half full of corn.

"Feed it to birds or squirrels in your back yard," I smiled and added, "Or you can plant it."

"Plant it? What will it grow?" he asked.

Dad and I looked at each other dumbfounded. Dad developed that twinkley chuckle in the corner of his eye and spread his hands out as if to signify he wasn't going to cover this one.

"Tomatoes," I replied, to the question deadpan.

"Really? They will grow?" incredibly, the man said to us. Then to his son, "Come on, Son, we're going home to plant some tomatoes."

Sir, whomever you are, wherever you are, I am sincerely sorry, it had been a long day. This chapter is for you.

Chapter 88
Holiday Time

Hey, set yourself, pull up a chair,
The sounds of the Holiday are in the air.
Snow covers the ground here, pristine and white,
winter in the country is quite the sight.
The Family's High Holidays are here,
I cooked a big lunch for those hunting deer.
So it's time once again for my message to you,
of all my adventures in two thousand two.
The year started out with a pleasant surprise
When four little lions materialized.
Kept two, left them with the pride,
Wwho kept them nice and warm inside.
After all these years it 'twas 'bout time,
The lionesses found motherhood sublime.
The other two cubs went on their way,
I saw them on the morning news one day.
Let's see what happened at the zoo,
I'm always tap-dancing, that I can tell you.
Parker delivered a c-section roadkill mom fawn,
who's still prancing across the zoo lawn.
He came in sans umbilical cord, which is a problem,
Take my word.
I pushed together and tied with a string,
a weird outty belly button for him this spring.
So "The Professor" (who came from Rutgers Street),
is the only buck with an outty- neat!
Jackie's wish finally came true,
When I brought her a puppy home from the zoo.
Cute, fuzzy, brown and some red all over,

THE ZOOKEEPER'S DAUGHTER

Jack was happy (thought I'd snowed 'er).
But the genetic truth came out one day,
When the little coyote growled in play.
Raised more fawns, two possums, and two raccoons,
hoping for lions and llamas soon.
A new male llama, dumb virgin ya know?
Gets all "Dressed up" don't know where to go.
Remember Chester the tiny little dwarf buck?
Had to put him down, genetic bad luck.
I have a great fan club of boys under eight,
Where were all these cute guys when I needed a date?
Space Farms had a 75th anniversary day,
I made the newspaper with my Alligator Pate,
I felt so Martha Stewart - esque,
But it was so hot, didn't look my best.
Oh well, for me that is nothing new,
I get pretty funky working at the zoo.
One Sunday morning we had a surprise,
on the porch sat a cockatoo with deep black eyes.
"Angie" the note said, "is sweet as can be.
She eats fruits and veggies," Signed S.D.
They saw Ted and I on NJ TV
And then brought Angie home to me.
(Read that: Lori is a soft touch)
Angie dances when I sing,
Though Dad says the poor soul is just convulsing!
My cousin Dutch stopped by one day,
Angie chewed up his Holiday pay.
Serves him right, Dutch was mean,
Angie just wanted to be part of the scene.
He *was* making a lot of noise,
Trying to be friendly with the hunting boys.
Dutch slung his coat over her cage,
Angie pick pocketed, no reason for rage.
The bird was bored, found something to do,
now Dutch's Christmas will sure be blue.
The Bear Safety Coloring Book went NJ state wide,
Fills my heart with 4-H pride.

LORI SPACE DAY

A feather in our cap, I must say,
Perhaps it will go nationwide someday.
More articles written for the zoo trade,
Speeches & TV, my life's a parade.
I painted & stenciled the rest rooms more,
Dad can't see the need for feathers on the door.
A new honor bestowed especially upon me,
a tradition passed down in my strange family-
Dad taught me his procedure to de-scent a skunk,
after the first, yes we both stunk.
After the fifth, I just didn't care,
But Doug, for days, was sniffing the air.
How do I list THAT on my resume?
Never can tell, may need it someday.
Canada in the summer, our great fishing place,
Pittsburgh in the fall, a smile on my face.
Doug's doin' great he works hard all the time,
definitely the brains in this family of mine.
Quiet and strong, people don't read him quite true,
He's a cracker jack joker, makes me laugh when I'm blue.
Which happens more often than I'd like to admit,
he cracks me up with his sardonic wit.
Jack's a teenager, oh boy do I know,
"Drive me here, there, the mall, the picture show?"
She's writing now also the newspaper at school,
A feature writer - she is so cool.
Tall, blonde, beautiful, smart as can be,
She's a poet (apple and the tree?)
That's my year, another one at the zoo,
always chasing time with more things to do.
To all my friends where'ere you be,
Put a smile on your heart and think of me.
Write, call or visit, you are truly missed,
for now, consider yourself Holiday Kissed.
Ecclesiastes 3: 1 through 10,
Until I see you once again.
Love, Doug, Jack and Lori

Chapter 89
Goodness Gracious, Great Balls of Fur!

If the phone rings at my house before eight in the morning, I know something has happened, usually bad. Everyone knows I am not the best morning person and not to call that early. In the off season, when the zoo is closed and there are no babies, I work Monday through Friday. I take the weekends off to be with Jackie, going down to the zoo at my leisure to take care of certain animals that only I work with. And never call on Sunday before church.

On Sunday March 9th, at seven a.m., I was barely awake, contemplating getting ready for church. I had done my homework and the Children's Sermon was prepared for that day. The phone rang. I slugged to the phone with my trustworthy coffee cup in hand. It was Dad.

"Hi, you awake?"

"Well, sort of," I replied.

Dad's voice belied his excitement. "Doug and Parker just found three tiger cubs, we are bringing them in." I could hear zoo noises so I knew Dad was on his cell phone. Suddenly, I was wide awake.

"I'll be right there," I chuckled. Neglecting my morning routine of face washing, tooth brushing, etc., I threw on the first clothes I could find and was out the door. My mind was racing faster than my car as I covered the quarter mile to the zoo office.

There was snow on the ground, left over from a harsh winter, and the temperature was freezing. If the guys were bringing the cubs in so soon after birth, there must have been a problem, as we usually give a mom a chance. But Tara had had a miscarriage the fall before... I was well stocked on formula, had bottles out the wazzoo, lots of clean towels and five heating pads. I did not know what I would find.

I pulled into the office at the same time Dad did. I unlocked the office door as Dad carried in a cardboard box. My excitement was tempered with concern for the health of the cubs. Dad and I looked in the box and to my relief the three good sized cubs appeared to be full term. Dad explained that the weather was cold and Tara, the mom tiger, was not staying inside the warm den where she should be. Dad and I set to work tying off umbilical cords, weighing the cubs and giving preliminary examinations. The three cubs were in good shape. Doug and Parker were outside cleaning the den, putting down fresh hay as the birth had soiled the previous Friday's change of hay. Unexpectedly, Doug burst through the door, a grin hidden under his winter beard.

"Got two more, they were under the hay. They're real cold," he shivered.

Dad and I looked at each other; this report was not good. The first three cubs were fine, already sleeping on their heating pads. Dad and I immediately turned our attention to the cubs in Doug's grungy farm coat. Dad took one, I took one and we felt their lack of body warmth. The little cubs had ice cold ears, tiny tails and blue tongues. Tara must have had these cubs first and in her inexperience with cubs, buried them under the bedding hay; big cats like to bury their excretions, too. The cubsicles were lethargic, barely breathing and not making a sound. Not a good sign. I knew what to do without Dad having to remind me. To his credit he didn't even start. I told Dad to take mine and I ran to the kitchen to get Gramma's largest stew pot. I returned to the office with a large pot full of warm water. Dad was gently massaging the cold cubs as I took one from his arms and dipped it in the warm water, holding it's tiny head above the water with my thumb and forefinger. The inert cub was small, not moving and still covered in birth fluids that now swirled colorfully into the warm water. I swished water around him and massaged him with my free hand under the covering warmth of the bath. He opened his mouth and meowed without sound.

"That's a good sign, "Dad said. "Get him out and I'll dry him while you dip her." We switched cubs, Dad briskly rubbing his cub with a towel, while I dipped the tiny female into the stewpot bathtub. *If Gramma could see us now,* I thought, *she'd be proud. Not worried about the gooey mess in her favorite big stewpot, not my Gram, not in this family!* Thankfully the little girl responded just like her brother and started to move. Dad and I finished rubbing the last two cubs dry and tied off their umbilical cords.

"Good luck." Dad grinned. "Get 'em fed and see how those two do." At forty-eight years old, I knew what needed to be done but was grateful for his help.

I simply said, "Ok, Pop, thanks."

I whipped up the formula and each cub drank its allotted two ounces. Even the cold ones. This was good. I checked my watch and saw it was only eight-thirty, I still had time to make it to the Beemerville Church, which is next door to the zoo in downtown Beemerville. I loaded the five cubs into a clean cardboard box, with towels and heating pads heaped on top of them for warmth and headed up the hill to my house. I plugged in the heating pads, and did the quickest makeover in recorded history and was out the door to church after calling our vet to let him know that the cubs were born, and we may have a problem with two. I asked him to keep his cell phone on. It was a Sunday after all. Emergencies always happen on Sundays, don't they?

I had two hours till they needed to be fed again, but only two minutes till church. I gave the Children's Sermon, dying to tell my secret, but could not. Dad and I discussed the birth during our examinations and decided since opening day was only a few weeks away, and we were not sure if all the cubs would make it, we would hold off on telling anyone about them. What a secret to keep. My first tiger grandchildren, the first tigers I would ever rear from birth, and I could not say anything to anyone outside the family. For three whole weeks! I am a gabby person, my Dad will tell you, and my elation was hard to contain.

He needn't have worried, however, as my joy and elation very quickly turned to joy and exhaustion after ten days of round-the-clock, every-two-hour feedings for five hungry cubs. By the time I fed five cubs, burped five cubs, stimulated five cubs for urine and bowel movements and cleaned five cardboard nest boxes, I had one hour to spend before the next feeding. I was amazed at the difference between the Atlas African lion cubs I had raised in the past and this new litter of tiger cubs. I had received Khyber and Tara when they were four weeks old. I assumed their aggressiveness on the bottle and toward me was due to the fact that they were raised by the tiger mom and did not know me. After about a week, as soon as Tara and Khyber got to know me, they had settled right down. The lion cubs I had previously raised from birth, were marshmallows until they became aggressive on the bottle at seven weeks old. The same with cougar cubs. No one was more surprised than me at the aggressive behavior of the five tiny tiger cubs.

Rex and Cleopatra, named for their distinctive hieroglyphic type eye shaped markings on their tails, were the original cold babies. Born at 3 and 2 1/2 pounds respectively they were just as aggressive on the bottle right from the start. Chamomile was an easygoing lady on the bottle. One of the little boys who would eventually be named Rocky attacked the bottle with growls as a newborn. And Peter (who's real name was actually P.I.T.A.- as in pain in the ###!) was searching for the nipple of the bottle during his second feeding and could not find it. I was trying to get it in his mouth; he just could not find it. So what did that little three-pound whippersnapper do? In total frustration he turned his head away from the bottle and clamped his little jaw hard on my wrist, with a growl to boot! I was flabbergasted. There is a difference between nudging around for a nipple and a bite. He bit me. No doubt in my mind. He had no teeth but his intention was clear at less than twenty-four hours old. All five tiger cubs wanted to nurse in the head down position. African lion cubs nurse in a head up position, their natural mother lions have large, rounded bellies and the cubs must reach up for the nipple as mom lays down. Tigers are often called "slab sided" because they are so narrow from side to side. My educated guess says that nature has evolved so tiger cubs nurse in a head down position. So every feeding was a struggle for a couple of weeks until I trained them to nurse in a heads up position that allowed the formula to flow out of a hand held bottle (otherwise they would just suck air).

I used the same color coded nipple rings on the bottle and collar system that I had developed to keep the lion cubs straight. It made it easier when I had to work with them and keep records until I learned their individual markings and personalities. Peter (I could not call him P.I.T.A in public) was the easiest. He was the largest weighing in at 3 1/4 pounds at birth and the largest in body structure. Peter was also the only cub that cried every forty-five minutes to be fed. A true pain in the ###! I set the cubs up in the office nursery in the daytime. If Peter cried for food I could walk out and close the door. He would cuddle with his stuffed animal and go back to sleep. At home was another story. He would whine every forty-five minutes and I would try to let him cry. Then Doug or Jackie would complain and make me feel guilty. Doug and Jackie would, occasionally, really lay it on thick. So I will admit, I often broke schedule and gave Peter a "tidy over snack" of an extra ounce until the next feeding.

Cammy started at 3 pounds with Rex and Rocky. Cleo was the smallest at 2 1/2. All good sized cubs. After a few days the litter and I settled into routines and Cleo quickly caught up in weight. The risk of pneumonia for Rex and Cleo was great, they were so cold when they came in, and Dad and I kept an eye on their breathing. At one week I stopped worrying so much (or maybe I was just too tired), as I figured their time to develop pneumonia had elapsed.

At two weeks old they had all doubled their weight and *Thank God*, started sleeping through the night. At least from midnight till 6 a.m. I felt like a new woman the first morning that I did not wake up to their meows for bottles in the middle of the night. Every new parent knows exactly what I am talking about.

Tiger cubs are also much more mobile at an early age than African lion cubs, which was another surprise to me. By one week old the little striped cats were walking, by two weeks they were trotting. Tiger cub eyes do not open till day fourteen through seventeen so my living room looked like a large crib with baby bumpers around it. I had placed pillows all over the place so they did not bump into anything and hurt themselves due to their lack of coordination. Toddling tiger cubs can be quite funny and I must admit Doug, Jackie and I burst out laughing a couple of times as they toddled and crashed during exercise time in the evenings.

I was a little worried about the smallest male cub. He was lagging behind on his walking skills and never really toddled along like the other cubs. All newborn cubs (like humans) have what I call the newborn shakes. Immature nervous and muscle systems cause movements to be jerky. Most cubs grow out of this stage by two weeks old, as their systems mature with use. The smallest male cub lagged behind the others on his development. When he tried to walk at two weeks, he would shake and crash to the floor. We named him Rocky. Rocky would sit quietly, watching the other cubs as soon as his eyes opened. You could see him contemplate the concept of walking, then try and crash. At two weeks the other cubs were walking well and started to trot. Rocky had not conquered walking. But he could run! He would study the other cubs, slowly stand, shakily stretch, then take off in a dash, ending in a crash to the floor once again. Doug and I would watch him at the evening exercise time and start to hum the Rocky Balboa theme song when Rocky would first concentrate, then dash across the living room. I was concerned with his neurological development, so I increased his vitamin B's to equal that given to a 12-year-old child. Vitamin B complex is one of the building blocks of the neurological system; I wanted him to have the maximum, more than he needed for his little body. Extra vitamin B's are eliminated though urine, so what he didn't need he excreted. I don't know if it was the only cause of his recovery, but Vitamin B complex is the family remedy for any neurological problems.

At three weeks old, Dad and I decided it was time to release the birth announcements to the newspapers. It was one week till opening day and I could not contain my excitement any longer. All the local newspapers came to see the cubs and do articles on them. Unfortunately, my picture was often taken with the cubs, wrinkles and all. My looks may be fading, but my joy was obvious. You cannot have youth and experience at the same time. I would not trade this experience for the world.

My babies were doing fine. Rocky had improved and all had tripled their birth weight. That formula is good stuff! I also had to put the word out in the zoo newspapers that the cubs would soon be available. As much as I loved my new cubs, I knew they could not stay at the farm. If we put them back in with the parents, Khyber, the father would kill his unknown sons and eventually breed with the females. So off to other zoos and trainers was the only option. I knew that from the moment the cubs came in to me, but it did not make the call to the zoo newspapers any easier. You simply cannot care for an animal for twenty-four-seven and not get attached. Each cub had its own personality, its own markings on colorful coats of fuzz, and for me, its own ideal placement.

Speaking of colorful coats, the tiger cubs were all born with rusty *brown* fur with black and white stripes. That surprised me. They were adorable, looked more like stuffed animals than the real thing until they moved. The brown fur shed out and by three weeks they were slowly turning orange. I checked with other tiger breeding zoos and trainers who all agreed that this was normal. Learn something new every day.

Rex and Cleo were the first cubs to be picked up by a trainer from Texas. I had dealt with this trainer before; he had one of my lion cubs. The trainer had the lion cub, which was now full grown, riding horseback. Now that tells me two things: the trainer was a good horse trainer, and more importantly, that lion never went hungry! I was sad, but ultimately somewhat relieved, to see Rex and Cleo go. Sad to have them leave me, but relieved of 2/5 mouths on the bottles, time, potty schedule and the massive amounts of towel laundry.

Peter, Cammy and Rocky were still coming home every night for their exercise time and bottles late at night and early in the morning. Living in large computer boxes I had garnered from the local bank, they progressed in development, weight and love in my heart. Dr. Ted Spinks and his entire animal hospital staff loved it when the cubs came to visit for shots and check ups. Other pet owners in the waiting room often were surprised when I showed up with the cubs. The pet clients would sniff and bark at the unusual cats. I dare say none of the pets had ever seen a tiger cub up close.

The tiny tigers were a big hit at the zoo nursery, visitors viewing them through the nursery office windows until they were four weeks old and about 12 pounds each. The cubs graduated to the Barnyard nursery outside and enjoyed frolicking in the grass all day, then snoozing in the shade. I gave speeches on the tigers twice a day to the crowd of visitors that gathered for their feedings. Often my niece Lindsey, now seven years old, would ask me for the key to the nursery. I would find her hours later still playing with the cubs while they were still "safe" for a child her size. Lindsey had memorized my tiger cub speech and would give her own version to the visitors that would gather. We were all very proud of her.

I was invited to take them to News12 NJ on *The Pet Stop*. Dr. Spinks and I had done the show before and we enjoyed the friendship of Dr. Voynick, the host. The cubs behaved well (Doesn't every mother worry about their children's behavior in public?) and chewed on the doctor's fingers as the cubs had started cutting teeth at three weeks old. I started the tigers on ground venison early as they were growing fast and their teeth were breaking though the gums earlier than the lion cubs had.

All the tiger cubs were aggressive on the bottle until about seven weeks old. Once again I was in store for a new learning experience. Just the opposite of the African lion cubs I had reared, the tigers suddenly became more mellow on the bottle. What a pleasant surprise! The tiger cubs that had asserted their "right" to the bottle, claws extended grabbing onto my arm with such audacity, calmed down. The cubs became cuddly and docile during feedings. The unfed cubs patiently waited, rubbing on my legs like a house cat while I fed them one at a time. I had worked with the cubs, rolling them on their backs, rubbing tummies and scratching under chins (which was their favorite spot). They had known no other mother, so there was no reason for their assertive behavior on the bottle other than instinct. I am constantly amazed by the difference between the species.

At seven weeks old the cubs were running and jumping all over my house, the curtains had been pinned up two weeks before. The weather had warmed and I decided to let the cubs try (or actually, let me try letting them) to sleep in the zoo overnight. Cuddle toys and towels were placed in their den. The cubs did fine. So did I after I checked on them twice, once at nine p.m. with bottle and again at eleven p.m. just to see if they were sleeping.

Peter was still very aggressive on the bottle and in his play. I knew he would need to be a zoo cat, you can never trust a tiger, never turn your back on them, and Peter was all tiger. Cammy's personality was calmer, but she was Peter's playmate from the start. I hoped I could place them together. I wanted a special home for Rocky, someone that would watch over him carefully. I always root for the underdog. Because of his shaky start, he was given and gave a lot more affection than the others. I held off till I found the perfect home for Rocky. Rocky went home with a trainer based out of Las Vegas. The trainer had experience and two little girls that would help heap on the love I felt Rocky deserved. Rocky left me at eight weeks old.

 I had put the word out to other zoos that I had a pair of cubs for adoption. Zoo people talk just like in any other business, and we have a reputation for strong, healthy animals. I was contacted by a new zoo that was just starting up. They wanted a pair of cubs, but I did not know the new zoo director. I called around and heard good things about him, so I agreed to let him come see the cubs.

 John Roberts showed up at the zoo with a vehicle full of assorted animals, after closing time. He was on a collection trip, bringing animals back to the Tri State Zoological Park in Maryland. After a quick tour of our zoo I introduced John to the parents of the cubs. Then we headed back to the nursery. I let the now twelve-week-old pair of ruffians loose in the barnyard to play with John (and the chickens, of which the cubs had not caught any of, yet). Cammy and Peter were hesitant at first. The suddenly timid cubs had not played with anyone outside the family, with the exception of Karen, our zoo photographer. I watched John and his interaction with my babies closely, explaining the training that I had started. John was knowledgeable, warm, playful and more importantly, loving with my two remaining babies. As we chatted he told me that he would not keep both cubs, but would trade one for an unrelated mate after a few months. I knew that should eventually happen no matter if the cats were living in the wild or captivity. After about an hour, I decided John would take good care of my special charges. I let John know that I would let him take the cubs with him. John looked at me in surprise. I don't think he realized that I had not decided whether or not he would be taking the cubs before he made the five-hour trip to our zoo. We went inside to complete the paper work. I gave John the bottles the cubs were used to using and a good supply of formula along with a written diet.

 After all the playtime, paper work, and instructions, I could not put the cubs departure off any longer. Jon had a long trip ahead of him and it was starting to get dark. John pulled his vehicle into the zoo and we put the two twelve-week-old, twenty-five pounds of claws and cuddly fluff into separate pet carriers in the back of his truck. John assured me I had visitation rights. I said my goodbyes; the cover of impending darkness hid my brimming eyes.

THE ZOOKEEPER'S DAUGHTER

I didn't want the new zoo director to think I was not professional, but I'm sure he knew. And so did Doug and Jackie when I got home. My heart sunk to my toes on my tear-streaked sneakers. My first tiger grandbabies were gone, off to a new life—without me.

Tiger cubs are lots of fun when they are small,
but quickly grow into a rambunctious hand full.
Cammy, at five weeks old, loved to play peek-a-boo.
Photo credit: Karen Talasco

Chapter 90
Angie Can Fly!

Every spring the northern region Girl Scout Camp presents a program for up to four-hundred-fifty Girl Scouts called, "Animals, Animals, Animals." I was a Girl Scout as a kid and enjoyed it immensely. So I like to help out where I can. I am one part of a five-part program. I speak to thirty to fifty girls at a time about my career, possible jobs for them as they get older and whatever animals I bring. It is a long day but very rewarding to talk to the young girls as they are amazed by my animals and my job.

By May the tiger cubs were much too frisky to take on an outing of this kind. So I packed up two baby coons, my corn snakes and Angie, since we had discovered he was so friendly. All were nestled in the back of my SUV. The Girl Scout camp is about forty-five minutes away from the zoo. I cranked up the tunes as Angie likes to sing with me and has also learned to imitate my laugh. I was about fifteen minutes away from the zoo just outside the town of Sussex, when I glimpsed a white crow flying out of the corner of my eye. *Wait a minute there are no white crows, what was that?* I looked to the back of the car and there was Angie, my cockatoo sitting on the back seat. *Well! What do I do now? If I stop and open the door he may fly away.* I pressed the window button to put all the windows up. Angie seemed content sitting on the back seat so I traveled on. I could do a half-hour trip with the bird in the back seat. Sure I could.

Around the next corner and Angie flew/hopped to the passenger's seat headrest. Ok, new game plan, but I can drive with a cockatoo on the headrest, why not? Barretta did it all the time! It's only a half hour. Umm, Hummm, let's sing some tunes and keep Angie calm. "My Boyfriend's Back" was on the radio and I knew every word. Evidently so did Angie. Angie was happy on his highest perch in the car. He spread out his salmon head crest and was bopping up and down to the music. I prayed a cop would not stop me, as I would not be able to roll the window down to talk to him. My dancing bird loved his position continuing his antics for another mile. People passed me head on and I could see the wide-eyed look of the drivers and passengers. Things were fine, I was singing at the top of my lungs with one eye on my escapee. Then the song ended and the DJ had the nerve to start talking!

Angie decided this would be a good time to express his affection for me and started to bob up and down with a clucking sound. I knew what that sound was, Angie was throwing me kisses. *Ok only twenty-five minutes to go, the next song would be on soon.* So I clucked back to my loving avian friend. *Wrong move, Lor, wrong move!* Angie decided he needed more than just the sound of kisses and flapped over to my driver's side headrest. Still making kissing noises, he pushed his beak into my hair and climbed down to sit on my shoulder. Curvy country roads and a large black beak too close to my ears and eyes do not mix. Angie had stopped kissing and was now preening me, looking for bugs in my hair and ears. I looked for a place to pull over and spotted one not too far ahead. I had never seen Angie fly when he was out in the main building; he had always stayed on my shoulder or arm. I figured I had a chance to calmly take him to the back of the car, open the hatch and get him back in his cage. The road I was on was not well traveled and dotted with the occasional house. If I could just stay calm, I had a chance. Besides, Angie can't fly. I was certain his wing muscles had atrophied due to lack of flight exercise.

I put the car in park and pushed the release button for the rear hatch. I slowly opened my driver's side door. Moving as steady and graceful as any ballerina, not wanting to jar Angie or frighten him, I slowly walked to the back of the car. I opened the hatch, so far, so good.

"BEEEEEEP!" suddenly filled the air. One of my friends drove past me in a truck and beeped the horn. *Just my luck.* Angie jumped off my shoulder and coasted down the hill toward an empty field. In slow motion horror I saw Angie's instinct take hold as he lifted and flapped his wings in flight that had been denied him so long as a captive bird. Fifty feet, a hundred feet, two hundred feet and half a dozen wing flaps later, Angie came to rest on the other side of a stream in the vacant field. Sitting on the ground, looking at the strange stuff called grass. I grabbed the cage, climbed through a barbed wire fence and ran toward the stream. The water was cold on my sneakers as I sloshed through the three-feet-wide stream. Not enough water to cover my socks, just enough to get my feet really wet and cold. Angie was watching me, his salmon crest erect, head cocked to the side and his wings limp in exhaustion. I was hoping that he was afraid of his new surroundings and would come back to me readily. I forded the stream and placed the cage next to Angie. As I had hoped, Angie was glad to see his familiar home and climbed right on top of the cage. I was taking no chances again and coached Angie back onto my arm and then placed him inside the cage. Fording the stream once more I carried the caged parrot back to the car. I searched the car for anything to shut the feed cup door Angie had escaped through. One paperclip did the job on one side and the calf clip on my zoo key ring did the other. Angie was safe. *Whew!* I took off my sneakers and socks, and did the centrifugal force dance with my sneakers hoping to wring out what water I could. I drove barefoot the last forty minutes of my trip and put on soggy sneakers sans gooey socks upon my arrival at the camp. The rest of the day went fine; Angie delighted the girls with his antics. He was excited from his flight, I am sure the same adrenaline pumped through him as through me. I spoke for the whole program, five hours worth, with wet squishy sneakers, but nobody knew but me. And Angie... he looked at my noisy sneakers and laughed right out loud... all day.

Chapter 91
Uh Oh, Snake's Loose

Traveling with animals in a car is never easy. Part of my public relations responsibilities is to go to remote locations (anywhere outside the zoo) and set up our Space Farms booth with pictures, brochures and assorted animals, depending upon what animals are transportable (read that small enough) at the time. Dr. Voynick, from *The Pet Stop* on News 12 NJ, was having an open house at his American Animal Hospital in Randolph, NJ. He asked if I could come and bring some babies. Sure, no problem at all, sounds like fun. June was rainy and it was pouring rain on the appointed Saturday. I got up early to complete my zoo chores before starting out for the day. Doug helped me pack up my car with a white tail fawn, two baby raccoons and my favorite snakes of the year, two red rat snakes, Flame and Lenny. While we were packing up, Doug said, "Aren't you going to bag up those snakes?" I explained that they were going to be viewed all day so I would just take the whole glass tank. A kiss on the cheek and I was off. I arrived at the American Animal Hospital in Randolph about ten a.m. and set up my area in an examination room. The day went well; there were lots of very interesting educational activities. I enjoyed the camaraderie of other animal-oriented people, vets and vet techs. I learned a lot as I visited all the other rooms while vet techs watched over my critters. At four p.m. the day was over and I packed up to come home. It was a busy time at the hospital, as all the other exhibitors were packing up also. The vet techs were very friendly and helpful, carrying out animals and other booth paraphernalia. I checked to make sure in the confusion I had not left anything or anyone, animals included. I had the tank of snakes, the fawn, raccoons and bottles, the important stuff and all the Space Farms promotional materials I had leftover. I departed in the rain and drove west across Route 10, which is a notoriously bumpy road, as I found out later.

Arriving home just before the zoo closed, Doug helped me unload the SUV and get the babies settled in. I fed the last night round of bottles to the fawns and tigers I had left at the zoo. Doug unloaded the snake tank. Doug came to find me and asked, "Exactly how many snakes did you take?"

Uh oh, this is not a good question to hear. "I took two; they should be in there." THEY were not. Lenny, the one lonely snake peered at me through the glass tank. *Gees, oh man, he must have got out at the American Animal Hospital.*

"The clasp on the tank was loose," Doug informed me.

Great, there are a lot of people that are afraid of snakes. I could just see Dr. Voynick examining an animal, the snake slithering out from under a cabinet, and his client's owner passing out cold. Dr. Voynick also had an assistant that was really afraid of snakes. Guess I better eat humble pie and call him to let him know about the possibilities. I called Dr. Voynick to let him know the situation. He was very understanding and said he'd have his staff on the look out. I searched the SUV upside down and could not find the colorful, yet elusive, reptile.

Days passed. I did not hear a word from Dr. Voynick or his hospital staff. Quite honestly I figured Flame escaped the tank and slithered outside either from the hospital or from my SUV. My mechanical husband asked me daily if I had found the snake. A whole week later I jumped in my car to an awful smell. I knew that smell; I had smelled it many times before. It was not an odor you would forget. Acrid, nasty, snake poop! I searched the SUV and found the offending offering behind the rear seat fold down. Grabbing my favorite cleaning fluid, I wiped up the mess. I knew what I had to do next; I had to tell my mechanically gifted husband that the snake was loose in our SUV. I knew he would hit the roof, that car was his baby, you know how men can be about a piece of metal with wheels. I sought out Doug to tell him the news. I tried to make it cheery by saying, "The snake is alive." Doug's naturally dark countenance clouded over even more.

"We'll have to tear the car apart, if that snake dies in the hot sun the resale value of our SUV will be nil." We both knew the stench of dead snake in the sun and it would be worse in an enclosed space.

Over lunch that day Dad, Doug and I discussed the possibilities of catching the snake.

"I could put a mouse on a fish line," I volunteered, racking my brain with hope.

"Won't work," Dad said, "Flame may twist up in the fish line in some part of the car you can't get to."

Ok, made sense.

"Put the mouse in a cage," Dad suggested, "just big enough for the snake to get into, he'd eat the mouse and have a lump in his tummy and not be able to get out of the cage."

We hashed out the idea and decided that it wouldn't work either, as any hole big enough for the snake to get into was big enough for the tiny mouse to escape from. Then I had a brainstorm. Same theory, different animal. That night after work I set up a mink transport cage with one by one wire in the back of the car. Inside the cage I put one of the small black farm chicks. I knew the snake was hungry after passing all that smelly poop. The snakes had eaten chicks before and the chick was too big in diameter to escape though the one by one wire. I drove our SUV home and parked it under the window to the living room. The sacrificial chick was chirping his head off. That was part of the plan, when the chick stopped chirping I would go and check the car. Shortly after dusk, the fuzzy chick became quiet. I ran out to the car to look with a snake bag in hand. Flame was nowhere to be seen. The chick had simply fallen asleep. I shook the cage to wake him up and went back inside to wait. Again the chirping stopped and I checked again with the same result. *Oh, all right, I'll give it till morning.*

Doug's normally quiet, but funny, personality was souring. In the morning Doug and I checked the car at daybreak. "I'll have my tools ready after work," he growled, as we looked in at the lonely chick. The chick had earned its reprieve and I gave it back to its mom when I went down to the zoo that morning. It was a long day; Doug's menacing scowl was soon going to be a permanent facial expression if you believe the old wives' tale of a face freezing in position. After work that day Doug and I dismantled the entire back of our SUV, right down to the bare metal sides. Curse words in combinations I had never heard before peppered the air. Carpeting, plastic boarders and everything else that could possibly be dismantled came out of the car as we searched for the mysteriously vanished five-feet-long snake. We did not find my hidden friend. We searched every crevice that was an inch in diameter or more. Doug was livid. We taped the light switches and left the doors open, hoping the snake would escape. With lingering doubt, we did not put the car back together.

The next day the tiger cubs were due for a check up. I always try to get the first appointment in the morning so we are in and out fast. I packed the two remaining cubs into their large pet carrier, placing the now huge twenty-five-pound fluff balls in the back of the rattley striped carcass of our SUV. Dr. Spinks' office is only about twenty minutes away. I scanned the car for any signs of the snake but did not see him. I arrived at the Animal Hospital of Sussex County and parked right in front of the glass entrance door. Gathering the full baby bottles I had placed in the passenger seat (a full tummy makes the cubs easier to handle), I opened the driver's door and stepped out of the car. As I slammed the door shut I glimpsed the orangey red coloring of my missing snake! Flame was traveling from the front of the car under the driver's seat to the back of the car. I was barely able to grab the last six inches of his tail, he was moving so fast. I'm right handed so I had grabbed his tail with my right hand. I held on tight, the sucker was not going to get away, he was not going to die in my car. The snake had a different idea. He struggled and pulled against my hand, the rest of his five feet, two inch diameter body snaking around and under the back seat. Flame pulled my hand flush against the seat. I was not letting go. I was more concerned with keeping my marriage alive than Flame's present comfort level.

Doug's angry voice echoed in my ears, "If the snake dies, the smell would make the resale value nil." *Darn this snake is strong.* Holding the snake tail in my right hand I could not open the rear door to get a better grip on the snake's upper body. The wily reptile was now stretched between the front seat and the back seat with two thirds of his cylindrical body seeking shelter under the back seat of the car. His muscles were leveraged in the metal structures beneath the seats. *What to do, what to do?* I knew I would get bitten, no big deal, I had been bitten before. I had to get that snake out of the car.

The parking lot was deserted; no other clients were there yet. I looked through the large glass window and saw Joan Kiely, Dr. Spinks' receptionist. I laid on the horn of my car waving my one free hand frantically in a "come here" motion. She spied my antics and *Thank God!* came out the door to help. I explained the problem to Joan and she quickly opened the rear door. Reaching between the doorframe and the frame of the SUV I grabbed the body of the snake with my left hand. Having more snake in my hand I was able to walk around the door itself, the serpent still trying to seek shelter under the backseat. I gently tugged, exerting constant pressure for the snake to back out. Flame was not budging. I heard the horrible sound of Flame's back popping, just like when Doug gives me my favorite back crack. His scales rubbing backward against the metal under the seat gave him the advantage of a better grip than I had. But I was not going to let go. I anticipated at any moment the wayward reptile's body would snap in half. I didn't want that to happen, but I was not going to let go. I walked my hands farther up

Flame's body. Suddenly he wrapped his tail and exposed body around my wrist letting go of his grip on the underside of the seat. Flame slid out easily. I had him. I was ecstatic, I was not going to get divorced, I saved the car, the silly snake was coming out in one piece and my husband was going to be so happy. The ribbing I had taken from my Dad and co-workers would stop. The world was a wonderful place once again. The colorful snake was wrapped safely around my arm. I grabbed Joan, giving her a big bear hug as I did my happy dance in the empty parking lot. I'm sure we looked like a religious snake revival dance in the parking lot, me jumping up and down, whooping for joy hugging Joan with a five feet bright red and orange snake wrapped around my fore arm. Tears of joy brimmed my eyes as I examined Flame. My relief was tangible AND Flame was alive and apparently unhurt!

Joan handed me the snake bag I had kept in the car in the optimistic hope that I might possibly catch the snake. Flame went into the familiar snake bag readily. I double knotted the bag, well, just because.

Still needing to attend to the business at hand, we brought the tiger cubs and the captured snake into the office. The other vet techs had heard the commotion; Joan and I laughed through a quick rendition of the story. Joan was the hero of the day in my book. The cubs were amazingly quiet and well behaved for Dr. Spinks. After checking the cubs, we looked over Flame. He was fine, I had not broken his back and he has been ok ever since.

The drive back to the zoo was a happy one; I sang every song on the CD to the cubs and snake. Flame rested snuggly in his bag on the front passenger seat. As I pulled into Beemerville I saw Dad and Doug working outside the main building of the zoo. I grabbed the snake bag with Flame in it and held it outside the driver's window, tooting my horn. I got a thumbs up from Doug. I was grateful for my absolution from sin. It took Doug and I two hours to put the car back together that evening. We went out to dinner to celebrate. When we returned I checked on Flame. Flame had eaten his two mice with gusto, also.

Chapter 92
Nine O' Muck and Mire

The ladies at the admissions counter handle the phones on the weekends. The phones constantly ring. Potential visitors wanting information or directions, reservations for groups or birthday parties, animal information and the all important dead deer road locations are the main purposes of the phone calls. With the invention of cell phones, communication is constant. Once we even got a call from the front of the lions' enclosure inside the zoo, with the caller wanting information: "What time are the lions going to be fed?" When the phone rings we all look toward the admission counter to see who the call is for.

On a busy Sunday I help out at the two registers located at admissions, answering phones, handing out maps and answering visitor questions between speeches in the zoo and animal infant feedings. Fall's busy times are on the weekends. The local forest foliage brings brightly colored visitors to view the competing natural vistas.

September was warm and sunny in contrast to the rainy spring. I was manning the admissions desk in between tiger cub feedings from Tara's surprise second litter born in July. The phone rang. The man on the other end of the call asked me if we would take in a litter of possums from a road killed mom. Figuring that the litter was spring born and the litter would be almost half grown, about eight inches long, I explained that I would take them and release them in our woods, but not in the zoo.

"Good enough," the local gentleman said, "I'll bring them in in a little bit."

Having not asked for a more specific time, I hung up and continued working. The day passed and I must admit, I forgot about the call. Four hours later, a jean clad middle-aged man came toward the admissions desk with a large two-feet-by-two-feet cardboard box covered by a towel. It occurred to me that this was my opossum call. I assumed he had the towel on top to keep the nine squirmy eight half grown youngsters in. Possums can be fast when they want to. He asked for me at the desk and I introduced myself as I lifted my hand to the edge of the towel.

"You don't want to do that here," he said quickly, glancing around as he moved the box away from my hand.

"Ok, let's go outside," I simply replied. A quick glance in Alice and Daisy's direction let me know that they saw what was happening and knew that I was no longer available to help them.

We placed the makeshift cardboard nursery outside on a picnic table. I reached for the edge of the towel once again and received a nod from the possums' rescuer. Slowly I pealed back the towel so the possums could not escape through the tiny triangle of entering light. There were no possums to be seen so I lifted the towel a little more.

"Holy Cow! You brought me the dead mom, are the babies in another box?" I said, grossed out by the sight. The gaping, tooth filled mouth of the mother had been crushed, eyes protruding, staring at me in obvious death. Brain matter, fur and muscle tissue were barely distinguishable from one another, the head flattened like a pancake. Did I mention the flies and goo? The gruesome sight had me momentarily transfixed, rubbernecking at a long since passed accident.

"No, there are nine in there, still in her pouch," he calmly stated.

Still in her pouch? Then it dawned on me, this must be her second litter this year. Possums are born in the spring; if this old momma had lost her first litter, she would automatically go into heat, mate again, conceive and produce another litter. Mother Nature wastes no time.

I took off the entire towel, Mom wasn't going anywhere. I examined her belly and the slight movement of me touching her belly produced a wiggling of activity. Four babies the size of a spool of thread literally fell out of the pouch, still sucking on the long exhausted milk supply through the dead mom's rubber band looking nipples. EEEWwwy! The slime of her crushing death, combined with the death released bodily fluids, the flies, road gravel and dirty, uncleaned babies was a little much for me. Did I mention the maggots? I seldom used latex gloves, but this seemed like a good occasion. I dashed inside to the first aid kit and garnered a pair of latex gloves, returning to the oozing mass of mom and babies. I pulled the first four off the nipples and their sheer tenacity surprised me. The babies did not want to let go. The nipples stretched and then snapped back into the pouch like overextended rubber bands. Prying with two fingers, I manipulated five more young from the dead mother's pouch. Each baby was in a life and death struggle to stay on the empty dead nipple.

"How long has she been dead?" I winced, as I thought of the babies sucking dead milk.

"I saw her on my way to town and called you. When you said you'd take them, I picked them up on the way back; it's been about three hours."

Nature is amazing; these babies were not going to give up.

We left the mom in the cardboard box and put her in the garbage shed. The guys would pick her up and know what to do. I recruited the possum rescuer and after checking to see that the restaurant was closed, dragged him into the kitchen to assist me. The restaurant had the only warm water in the zoo. These babies were cold and dirty, covered in a mix of fluids that only God could differentiate. One by one I washed and he dried the maggoty, yucky babies. They were wiggling in response to the warmth of the water and I was careful not to get soap in their eyes. Ten minutes later seven healthy babies and two questionable babies were nestled in clean, fresh toweling on a heating pad in a new cardboard box marked "Beanie Babies." I moved them into the office nursery. I had a good supply of premature baby formula. I used an eyedropper to give them formula every couple of hours and let them sleep it off.

The next morning, one of the questionable two was dead. Its littermates had tumbled it out of the towel nest. Survival of the fittest. Mom possums in the wild lose a lot of their babies. The second questionable baby died in a couple of days. At least he died with a full tummy. The other seven progressed nicely, gaining weight, growing steadily. The large, silver, mouse-like creatures would greet my voice with shiny, beady eyes and a gaping, mini-toothed grin. I started them on a mixture of baby food and formula. I was greatly relieved when I observed the last hold out licking out of the dish. They would make it. Four weeks later they were all six inches long with long, clean, pink prehensile tails. I did not name them, as their markings were so similar I could not tell the difference anyway. Each had a black polka-dot on each white, paper-thin ear. Three males and four females. I often took them out for speeches, letting all seven cling with their tiny claws to my shirt as I fed them applesauce from a rubber-coated baby spoon. This was a natural instinct for them. In the wild they would ride on mom's back wherever she went.

We, for sure, did not need seven possums in the zoo. Due to their late-in-the-year birth, these babies would have died in the winter as there was no way mom could have supplied enough milk to feed them over the winter.

THE ZOOKEEPER'S DAUGHTER

They would be too small to survive on their own. The remaining babies stayed in the nursery office, and as the weather was getting colder outside in October, I planned on moving them directly to the animal room in the winter quarters barn. One day I walked into the nursery/stockroom to be greeted by scurrying feet. Three live possums were dashing in and out of the souvenir stuffed animals in the adjacent room. Their troll like features were camouflage amongst the stuffed animals, as long as they stood still. It was a Wednesday afternoon so I enlisted the help of my twelve 4-Hers to ferret out the posing possums. What a time we had laughing and tracking the fast moving critters in between boxes of fake animals. We found and captured two escapees, returning them to the nursery. I put a lid on the incubator. I set up a cage in the warm animal room in the winter barn. I planned to let them go in the spring when the weather broke.

Possums are too dumb to come out of the cold, and often have frostbitten toes, ears and tails. Their brains are slightly larger than a pea. They don't make a nest and travel constantly with no intentional direction. They eat only what they can scavenge. They walked the earth with the dinosaurs. You would never think the slow-witted possum could out maneuver twelve energetic pre-teen kids, and outsmart me. Nature is amazing. We never found the last escaped possum.

Chapter 93
G.I. Gator

Every state has a different set of laws governing their native fish, native wildlife and exotic animals that reside in that state. Many states allow individuals to have exotic animals such as, believe it or not, tigers, alligators or lions along with the "normal" exotics of large parrots, snakes, ferrets, etc. The State of New Jersey doesn't allow individuals to keep exotic large animals without a lot of red tape.

In the fall of 2002 we received a phone call from the distressed parents of a young fresh-out-of-boot-camp Marine. Their son was coming home from boot camp and had a small alligator.

"Can you take it in?" the father asked. I envisioned a one-foot-long gator; knowing that boot camp marines cannot keep pets, he must have picked it up on the way home.

"Where did it come from, " I asked.

"A pet shop," was the simple reply.

I knew the Marine would be in trouble with the state if they kept it. Dad and I had discussed putting an alligator and crocodile exhibit together as the zoo had had one years before.

"Let me call you back." I scurried to find a pencil and wrote down the number. I relayed the message to Dad and we decided to take the little gator in. We had a lot of fish tanks and would set him up in one until he grew large enough to go outside. Under good conditions an alligator baby will grow a foot a year, so I knew he would not be in the nursery long.

On the appointed day, the alligator's whole human family showed up to say goodbye. The young Marine stood tall, lean and buffed as he announced he would soon be deployed to the Middle East. The parents were understandably proud. The two-foot-long(!) gator was cradled in the eager Marine's arms. The gator's family also brought with them a seventy-gallon fish tank set up, the home of the small gator. I should have known to ask more questions on the phone, like, "How big is he now?" *I am not getting*

THE ZOOKEEPER'S DAUGHTER

any smarter with age, just a softer heart. We set the tank up in the main building. The G.I.'s gator was, at that time, very friendly and handleable. That was a good thing, as I have never seen myself as the leopard skin, bikini clad, alligator wrestling type. The young Marine said goodbye, sad to be leaving his pet, but proud to be serving his country.

G.I. Gator was happy in his tank, eating well, and spent his days basking under the heat lamp on his rock watching the visitors go by. Many a visitor was startled by G.I. Gator. The alligator would bask absolutely still, then for no apparent reason move his head or blink an eye. Visitors standing nearby would jump and scream. The main building has many stuffed animals in it. Some visitors would just assume Gator was one of them.

By the spring of 2003, the little gator had grown another foot and was getting crowded in the four-foot-long fish tank inside. I set G.I. Gator up in the outside nursery. He had a small section of grass with a kiddy pool in a section away from visitor hands. The fence surrounding his section was four feet high. The now three-foot gator enjoyed the sun, the grass, and would splash into his pool on a regular basis. Every morning I gave him fresh water in his pool and pulled up minnows from the minnow trap we kept in the lake bordering the nursery. He swam around and around in his kiddy pool, catching minnows with a showman's flair.

One morning I was shocked to see only the glistening sun in his pool. I searched the entire nursery yard, every clump of grass, every rock, every inch of fence line. G.I. Gator was nowhere to be seen. G.I. Gator had split. His natural camouflage worked well for him, G.I. Gator had gone undercover. I notified the rest of the keepers to keep a look out for the wandering reptile. We knew where he was, just next door in the waterfowl lake on the zoo property, we just did not see him. G.I. had found the perfect environment for an alligator. The waterfowl lake is a warm, spring-fed lake crowded with optimal alligator food: fish, turtles and, for desert, goslings or ducklings. A rowboat ride around the large pond on a weekly basis did not show any sign of the elusive alligator.

Fritzy, the black Australian Swan, had started a nest in the southern end of the lake. This was her first breeding season so I was visiting her daily to check on the number of eggs in the nest. One visit I saw a flash of movement out of the corner of my eye. I quickly scanned the area seeing nothing on first glance. The water surface was littered with tree branches, twigs and other debris the wind had blown in a southern direction. Then, slowly like a hidden picture puzzle, my eyes detected a familiar shape. G. I. Gator had surfaced five feet from shore near Fritzy's nest. His body completely submerged, only his eyes broke the surface of the water. He was looking straight at me so I froze in position. I studied him, happy to see G.I. alive. Not that I had any doubts, the lake was a perfect alligator environment, but we do have eagles and ospreys prey upon the fish in the large pond. I don't know, would a bird of prey take a two-and-a-half-foot alligator? I've seen them take two-feet koi out of the pond. Joe gator was not only alive, he was thriving in the lake. He had grown another half a foot; his belly was distended with food. His bubble eyes were clear and sharp. August marked two months G.I. Gator had been undercover in the lake.

Earlier that spring Dad, Doug and Parker had constructed an alligator and crocodile exhibit bordering the lake. The rented (yes you can rent them) gators and crocs would bask in the sun in the warm afternoons. Their enclosure was half water, half land. The half dozen six footers in the exhibit attracted a lot of "Oohs" and "Ahs" from visitors. The easiest way to tell the difference between the two species is by the shape of the snout. The American Alligator has a broad snout and all the teeth in its upper jaw overlap with those in the lower jaw. The crocodile has a pair of enlarged teeth in the lower jaw, which fit into a notch on each side of its snout. These teeth can be seen when the jaw is shut. We were feeding the crocs and gators venison. Occasionally they would catch a wayward fish that wandered into the water section of their enclosure. The splashing activity attracted visitors, who watched in wonder.

The alligator is indigenous in the southeast section of the United States where it was once hunted for its skin. It neared extinction and a conservation program was set up in the 1950s. The program worked so well that the alligator is now controlled by organized hunting. The alligator we now see as boots, purses or food on exotic menus are raised in farms in the south. Zoos can rent alligators and that is what Dad decided to do since we had no winter quarters for the crocs and gators. They require warm conditions to digest food, as they are both cold-blooded reptiles. In the wild they absorb the heat they need from the sun. So I was not concerned that G.I. Gator had taken up residence in the pond, I knew he had a varied natural food supply, and the summer was warm enough. However I had to catch him out before fall and colder temperatures.

I set up a five-feet square net under the water off shore where Gator was frequenting. With strings attached to the four corners of the net and strung to the guard fence of the pond, the theory was that he would grab the venison chunk I left as bait and I would pull the strings trapping him in the net. It was a great theory, but it never worked. After repeated tries, the gator had started to avoid the area. Ancestors of this gator had walked with the dinosaurs between 225 and 65 million years ago. His survival instincts had been honed to perfection over the summer. I could not catch him. I faced reality after discussing it with Dad. G.I. Gator would freeze to death with the onset of winter, or possibly hibernate in one of the warm springs feeding the pond. Well, that was right hopeful thinking, most likely he would freeze to death.

Months passed with various sightings reported of the young, growing alligator. G. I. Gator would bask on the center island, or on the dam far from shore. As soon as we approached in a rowboat, he dashed back into the underwater safety of the murky pond. Friends joked with me that I would be donning that leopard skin bikini and wrestling that gator in the murky lake after all! We had no ducklings or goslings following moms on the zoo pond that summer. We knew where they were.

Every fall after we close the zoo the race is on to get the animals bedded down for cold weather. After that goal is accomplished, we start what projects we can before the snow comes and puts the kibosh on outside projects. One of my projects every year is to clean up the nursery area. I repainted one enclosure and tidied up the yard, putting away the kiddy pools, taking in the shovels, rakes, etc. until next spring. I was working in the corner of the Barnyard Nursery near the lake that borders it. There was branch and twig debris under the large willow tree from the last windy rainstorm we had. The sun speckled off the water, but the water temperature was cold. *But not today.* I made a mental note that the debris would need to be cleaned up next spring. As I glanced again across the fencing my eyes adjusted to the broken branches on the shore. *That is a strange log.* Low and behold, it was G.I. Gator. I had not seen him for two months. I figured him for a goner since the first cold snap weeks before. His dark brown mottled scales glistened in the sun. Alas, Gator was as still as death. His eyes were closed, the nictating membranes covering his eyes were cream colored. From a distance the eye coverings looked like dried out eye sockets of the desiccated corpses I had seen in Florida gift shops. I pulled out my keys to unlock the gate that led to the pond. Sunlight reflected off my zoo keys and I heard a commotion of splash. I looked up to find Gator gone once again. *He was alive? I really need to get my eyes checked.* I slowly waked to the water's edge to see if I could see him or if the young gator had scooted to the other end of the acre lake again. Periscoped eyes bloomed on the cold clear water. Wedged, protected and camouflaged by the floating branches and twigs, G.I. Gator silently observed me as I did him. Knowing that he was faster than me in the water, I slowly backed away.

I found Dad where he usually was this time of day, skinning dead road kill deer in the feed house. I grabbed the new snare Parker had recently purchased. It hung on the wall with other equipment. I gave Dad a quick rendition of my objective. The snare was white, shiny and new. Dad instructed me on its use and wished me luck. I hurried back to the pond, sneaking through a different gate, attempting to hide behind the trees and weeds along the way like a spy. G.I. Gator was in the same spot underwater just offshore. I readied my noose on the end of the six-feet PVC pole. I swung the noose toward his head. The sun reflected off the shiny cable and G.I. Gator silently submerged undercover once more. I froze in position, knowing he could see movement, but not color. Surfacing to my left, large beady eyes watched me, just out of reach. Dad had come across the street to observe and help.

"He can see the noose and the sunlight reflecting on the bright white pole. You have a better chance with the old 12-feet copper pipe noose, it will blend better." Dad stood there chuckling in his torn and bloody farmer's bibs. I tried one more time and this time the outsmarting gator disappeared completely from view.

"All right, I'll try after lunch," I replied. "I hope he comes back."

"He will, if he knows what is good for him!" Dad was still chuckling over his 49-year-old daughter alligator hunting in the Beemerville zoo pond. We left for lunch.

I cut lunch short. I had an objective that I could not postpone. The twelve-feet copper pipe snare was unwieldy in my hands. It did, however, give me two advantages to bring G.I. Gator in from the cold. The length would help me reach farther out into the clear, cold water. Its patina coloring did not reflect the sunlight and alert the wary alligator. The old willow tree was my cover again. Silently standing beside its weathered bark, I gazed into the water looking for G.I. Gator. I spied him under the branches and twigs just four feet from shore. He hadn't seen me yet. I waited with anxious patience, watching him, watching his surroundings. G.I. slowly closed his eyes. I moved my rusty wire snare over his head hovering above the water's surface. Alligator eyes opened again under the water. I froze in position. My arms ached from holding the uncomfortably cold, long copper pipe. I held my pose until Gator closed his eyes once more. Eyes closed for a moment, the undercover gator did not see the noose enter the chilly water. His eyes opened, my noose was inches away from his head.

THE ZOOKEEPER'S DAUGHTER

Patience, Lori, patience will do it. I heard Dad's words in my head. Amazingly the gator did not see the noose right in front of his nose. Dad was right, the rusty cable wire blended in color and texture with the small twigs in the water surrounding the hiding gator. Just because something is new does not make it better. A life lesson on a cold November day. The young gator rested his eyes closed one more time and I nudged the loop of the noose over the head of the submerged reptile. Like the child's game of Operation™, I tried to maneuver the loop past Gator's head without touching him. He opened his water filled eyes one more time and I knew it was now or never. I pulled the cable wire with all my might, the cable biting into my hand as I lifted the twelve-feet pipe like a heavy fishing pole. I had something in the noose, I hoped it was G.I. Gator, not just a branch. Swinging the snare pipe to dry land, I saw that I had the elusive critter. Put on dry land the reptile immediately tried to scoot back into the safety of the lake. The little gator struggled on the other end of the twelve-feet pole. He wanted to go back to the water and I wanted him to go in the other direction, farther from the water where I would have a chance to catch him after releasing Joe from the noose. Gaining the distance I felt I needed, I let go of the cable on my end of the copper pipe. G.I. Gator was struggling on the other end trying to free himself. The rusty wire noose did not open up. The cable was stuck. I needed help. I leapfrogged Gator on the end of the pole, not wanting to strangle him in the noose. We had to clear the pond area and climb six steps to the front yard of the main building. I knew the menfolk would be across the street after lunch. If I could just get us to the yard, I could holler loud and get some help. I kept leapfrogging G.I. Gator across the area to the steps. The steps were a challenge. I had to lift the four-feet gator ten feet in the air while climbing the steps. Finally reaching the top step I landed Gator on the grass on the other side of the four-feet guard fence surrounding the lake. G.I. Gator was on the other end like a pole vaulter, still struggling to escape the noose. I dragged the reluctant reptile to the front of the yard. I didn't want to lift him anymore, not knowing if the noose would injure his neck.

Dad's car was still in his driveway; he was still at lunch. I saw Doug and Parker driving from the feed house toward me. I hooted and hollered and swung my free arm with gusto, trying to flag them down for the extra hands I needed. Doug and Parker came our way. When they arrived I explained my predicament. Parker gently stepped on G.I.'s head while I loosened the noose cable from my end of the twelve-feet pole. Reaching down Parker nudged the stuck cable away from Gator's neck, which is by nature right next to his mouth full of slashing teeth. G.I. Gator was not happy, his mouth open and hissing. Doug helped me with the unwieldy, now bent copper pipe pole. Parker grasped the gator by the neck and handed him to me.

"He's cold," Parker said. Parker is a man of few words.

We all laughed. I grabbed the gator behind the neck, said, "Thanks, guys," and started walking up the hill toward Dad's basement where the fish tanks were kept. Dad came out of his house at that moment and saw me walking across the lawn with the captured, still struggling gator.

Dad gave me the thumbs up and hollered, "Congratulations!" to me.

We spent the next hour setting up Gator in a fish tank in the main building, the same tank that he was in the previous spring. The tank is a little small for him, especially after he ate his fill at the lake all summer long. But it is all we have to keep him in, inside where it is warm. G.I. Gator had grown a whole foot. He rounded out his body over the summer, stealthily snagging unsuspecting fish and fowl from the lake. Next spring he will go outside to the large alligator and crocodile enclosure. G.I. Gator came in from undercover in the cold water and now spends his winter days basking under the sun lamp being hand fed morsels. You can look in his protruding shifty eyes and read his thoughts. I know he is just waiting till spring for his next covert adventure.

Parker shows Hunter, his son, G.I. before G.I.'s great escape.
Photo credit: Karen Talasco

Chapter 94
Christmas Pearls

Oh My Gracious, gosh, golly gee,
It's only a week 'till Christmas of 2003!
This is my poem to friends true and dear,
Of all that has happened to us this past year:
Khyber and Tara finally came through
with five tiger cubs, three healthy, two blue.
Soaked two in Gram's stew kettle nice and warm,
They pulled through without lasting harm.
My spring was busy with five little cubs,
with bottles and burping and bellies to rub.
By June they were gone, alas my broken heart,
Off to new zoos, for them a new start.
Add two raccoons, three fawns, two possums and a gator
(More about that lousy lizard later).
Night speeches, day talks, questions at the zoo,
There's never a lack of things to do.
I've painted and cleaned till I'm blue in the face,
Keeping myself to a hurried pace.
Why you ask? I really don't know,
must be the challenge that keeps me on the go.
Helped Dad with new exotic snake dens,
Then painted the insides as was my yen.
Impressed myself with my brushstrokes fair,
But in truth, the snakes, they just don't care.
Two more tiger cubs, Tara you gem,
Boy do I love raising them.
Babies were placed, so long, goodbye.
I should be old enough not to cry.

THE ZOOKEEPER'S DAUGHTER

Then nine more possums I pulled from a pouch,
on their dead, tire-tracked mother (Ouch!)
And oh, what a yucky mess, what bodily fluids I can only guess.
Kept three, lost two, placed four, still keeping score?
Then one escaped to the office floor.
I've not found him yet, maybe he's dead,
The circle of life from a mom with a tire tracked head.
Oh and one more feather in my hat,
I received my first by-line, imagine that!
An article on the old Covered Wagon,
Printed in full, I'm proud, not braggin'.
Was on News 12 NJ three times–
With my friends Dr. Brian and Ted
Taking animals from the zoo, TV does stress my head,
You never know just what animals 'll do,
Spit, burp, fly, run, bite or poo!
Jack is amazing, now seventeen,
With a driving permit she's making the scene.
Chauffeuring Mom from here to there,
Every other week new color in her hair.
Her grades are great, must be from her dad,
Though on brains I ain't half bad!
This year I got old, or so Jackie says,
as she counted the gray hairs SHE put in my head.
"Get out of your box, Mom, you're in a rut!
Try this color?" All I could say was "But.. But.. But.."
So now she driving and I've gone crazy,
hope the new blonde doesn't make my brain lazy!
Fishing this summer, a wilderness dream,
face to face with a bear across the stream.
Saw otters, huge snapper and a loon,
resting all night under a picturesque moon.
Love our trip to Canada, a well deserved break
From the long days at the zoo and the pace that it makes.
Doug's still my hero, keeps my feet on the ground,
With plenty of laughter and jokes all around.
He's strong as an ox, somewhat scary to view,
but just a big teddy bear, intelligent too.

Just get him to smile and you will see,
The light of the world, according to me.
Our 4-H Coloring Book took fourth place,
in a national Colgate Palmolive showcase.
My kids were sure proud when the state printed the book,
It's now on the NJ Fish and Wildlife website- take a look.
An Alumni Award was bestowed on me,
From Rutgers 4-H in Sussex County.
Ah yes, and
The baby gator escaped early in the year,
to the pond at the zoo as I watched in fear,
Goslings, signets and ducklings disappeared.
Saw him all summer basking in the sun,
Soon as I approached in the water he'd run.
'Till finally this fall on a cold sunny day,
He basked on the shore not too far away.
In the water he slipped, slowly I placed my noose,
Right over his head, I "cooked his goose!"
Pulled him out onto the shore,
Then hollered for help, cause I sure needed more
Than my two hands busy on the pipe,
Or I'd suffer a God-awful bite.
Caught him by Golly, proud as can be,
This 'ole gal's still got it at nearly fifty!
Went to Florida to see Doug's family,
His folks are fine and sister Jackie.
Had Thanksgiving at a Chinese Buffet,
Then went to the beach for the rest of the day.
Every now and then I do something totally new,
Keeps the gray matter growing, a challenge too.
This fall I preached my first full fledged sermon,
"God's Time" was the title, now I know you're yearning-
To laugh right out loud - so just go right ahead,
God must of liked it, Lightning did not strike me dead.
The Beemerville pulpit will never be the same,
I admit *I* chuckled when *I* saw my name.
If all my old friends could just see me now,
they'd shake their heads and say, "Holy Cow!"

THE ZOOKEEPER'S DAUGHTER

Now friends, you know I am not all that prudish,
Just the only Beemerville Presbyterian Buddhist!
Alas, no reason to ramble on
For in the flash of an eyelash all is gone.
To my friends country wide I bid you adieu,
Drop me a line, tell me how you do.
Let me know what's happening in your world,
Inform this simple country girl
There is a life outside this cloister,
The pearl of wisdom in my oyster.
Come visit, I'm most always home,
Where the lions roar and the buffalo roam.
Though far away, not out of mind,
I wish you all the happiness you may find.
A friend of mine you'll always be,
Put a smile on your heart and think of me.
Love, Lori

Chapter 95
Dear Jackie

Dear Jackie,

You, Daddy and I have shared our lives with many animals, infants and adults alike. We have had adventures, heartaches and triumphs, peppered with love and laughter with our animal family. Treasure these experiences. There is nothing more rewarding than working with God's creatures and the wonderment of nature's creations. I know many times I have left you snug in your bed to tend to the animals that needed me. Daddy was there to watch over you. You were never alone.

You have done well in school and soon will be embarking on your college years. I have written this book for you to take with you. This is a book of our animal adventures, stories for you to read before bedtime. Just as years ago when I told you bedtime stories at camp of my animal friends at the zoo, I hope you doze off with happy, loving memories.

I cannot come with you to college. It is time for our little bird to fly the nest. So spread your wings and take that first flight. Never look back, but never forget either, migration is mandatory. Heritage is so much more important than inheritance. Gramma Lizzie told me that and it is true. You come from a long line of ancestors who love and care for more than themselves, caretakers of our earth. I know you will do us proud.

With this book under your pillow, you will have all our love with you, encouraging you to fly higher while you have soared from our sight. The best wishes of all our animal friends, both here and departed, will lighten your heart and guide your dreams.

I know we have taught you how to care, how to love and how to worry. Don't worry about Daddy and I, we will be fine. I will fill up my hours with work that is a joy to me and Daddy's days will be filled with the challenges that he encounters. I hope you can find the same joy and challenge in whatever you choose to do.

Never forget, but never look back with regret for what you should have done. Do everything to your best abilities and God will know. Work hard, play harder, and enjoy all that life has to offer. Time flies faster than a swift in free fall. Live the journey, not the destination.

Sleep well on your first night away from us. We will worry, but I know in my head you will be fine, my heart just needs a little more time to learn. I've had so much practice with the animal babies; I should know that by now. I am your mother and you are my only human child, so it is instinctual that I should worry.

...And Jackie, don't worry about that crazy lady with a flashlight standing outside your dorm window while you are sleeping. It's only me. I've cleared it with campus police.

Love, Mom.

Jackie and Peter the tiger cub.
Photo credit: Karen Talasco

Bibliography

Burton, Maurice & Burton, Robert. "Encyclopedia of the Animal Kingdom." London: Phoebus Publishing Co., 1976.

Cobon, John. "The Atlas of Snakes of the World." Neptune City, N.J.: T.F.H. Publications, 1991.

Hediger, H. "Wild Animals in Captivity: An Outline of the Biology of Zoological Gardens." New York: Dover Publica-tions Inc. 1964.

Hickman Sr., Cleveland P., Hickman Jr., Cleveland P. & Hickman, Frances M.. "Integrated Principles of Zoology." St Louis: CV Mosby Co., 1974.

Keeton, William T. "Biological Science." New York: WW Norton & Company, 1972.

Miller, Stephen A. & Harley, John P. "Zoology: The Animal Kingdom." Dubuque, IA, Bogata, Boston, Buenos Ares, Caracas, Chicago, Gilford, CT, London, Madrid, Mexico City, Sidney, Toronto, Wm C Brown Publishers, 1996.

National Geographic Society. "Wild Animals of North America." Washington, DC: National Geographic Society, 1987.

Rhodes, Dennis. "New Jersey Wildlife Illustrated." Union City, NJ: William H Wise & Co. Inc., 1977.

Rue III, Leonard Lee. "Pictorial Guide to the Mammals of North America." New York: Tomas Y Crowell Company, 1967

Rue III, Leonard Lee. Sportsman's "Guide to Game Animals: A Field Book of North American Species." New York, London: Outdoor Life Books,

1968.

Rue III, Leonard Lee. "Furbearing Animals of North America." New York: Crown Publishers, 1981.

Soucy Jr., Leonard J. & Decker, Carol. "New Jersey's Owls." Millington, NJ: Soucy Jr. & Decker, 1980.

Space, Fred T. "Facts About Snakes of the North East USA." Sussex, NJ: Fred T. Space, 1969.

Teitler, Risa. "Taming and Training Macaws." Neptune City, NJ: T.F.H. Publications Inc., 1979.

Again I must reiterate: Learn from the elder experts who surround you, and observation.

Printed in the United States
103737LV00007B/123/A